Chemical Information for Chemists
A Primer

D0073521

Chemical Information for Chemists
A Primer

Edited by

Judith N. Currano
University of Pennsylvania, USA
Email: currano@pobox.upenn.edu

and

Dana L. Roth
California Institute of Technology, Pasadena, USA
Email: dzrlib@library.caltech.edu

RSC Publishing

ISBN: 978-1-84973-551-3

A catalogue record for this book is available from the British Library

Published by The Royal Society of Chemistry,
Thomas Graham House, Science Park, Milton Road,
Cambridge CB4 0WF, UK

Registered Charity Number 207890

Visit our website at www.rsc.org/books

Preface

"The literature of chemistry is like a great, inspiring mountain with a core of rich ore ... in its use one must learn how to climb this mountain and must know where and how to dig for the ore ...."[1]

Chemical Information for Chemists: A Primer is written for students, scientists, and information professionals who will benefit from learning the tactics of finding chemical information. Although the title implies that the book is intended for chemists, we hope that all scientists who use or study chemicals will find it essential in developing an effective and productive approach to searching the complicated literature.

The structure of the book roughly follows the syllabus of a graduate-level course in chemical information offered at the University of Pennsylvania, beginning with the structure of the chemical literature and its use in the broader chemical research process and proceeding through various types of source and methods of using each effectively and efficiently. It is designed to be used either in a stepwise progression or as a collection of comprehensive reviews of specific topics and is roughly divided into three sections: an introduction to chemical information, the primary literature, and the secondary literature and specialized search techniques. Each chapter is authored by a recognized subject matter expert who brings both subject knowledge and search experience to the work.

The introductory material attempts to put the rest of the book in context, presenting information about the chemical publishing enterprise and discussing copyright rights and restrictions. The primary literature is addressed in two chapters. Chapter 2, Non-Patent Primary

Chemical Information for Chemists: A Primer
Edited by Judith N. Currano and Dana L. Roth
© The Royal Society of Chemistry 2014
Published by the Royal Society of Chemistry, www.rsc.org

Literature, presents the salient features of journal articles, conference papers, and reports, methods of locating and evaluating them, and tips for keeping up with current advances in your field. Chapter 3, Chemical Patents, presents a comprehensive overview of the patent as a legal and scientific document. The remainder of the book consists of chapters that teach practitioners to locate this primary literature, beginning with a chapter on text searching and the use (and misuse) of impact metrics and then moving into more specialized search techniques, including structure and substructure searching, finding properties of substances, identifying commercial sources and locating chemical safety and hazards information, polymer information retrieval, reaction searching, and performing sequence similarity searches using BLAST.

We hope that *Chemical Information for Chemists: A Primer* will aid you in developing sound strategies for retrieving data and literature related to your area of research by providing a clear explanation of search techniques and the underlying organization of the chemical literature; however, information retrieval is a very personalized skill. The more frequently you search for information, the more you will discover exactly which practices and procedures best suit your needs at each stage of your work. Conversation with information professionals and more experienced searchers can also help you to select an appropriate tool and strategy in a given situation. Chemical information is a science of trial and error; the more searches you try, the more applicable results you will retrieve, and the broader experience you will achieve. Before long, you will develop your own methods of finding what you need, allowing you to add information retrieval to the list of techniques you employ to succeed in your research.

Judith N. Currano
Dana L. Roth

REFERENCE

1. E. J. Crane, A. M. Patterson and E. B. Marr. *A Guide to the Literature of Chemistry*, 2nd edn, John Wiley & Sons, 1957, p. xiii.

Contents

Chemical Information for Chemists: A Primer
Edited by Judith N. Currano and Dana L. Roth
© The Royal Society of Chemistry 2014
Published by the Royal Society of Chemistry, www.rsc.org

II. The Primary Literature

I
Introduction

the library. The scholarly communication cycle is at the core of the scientific endeavor for both research and teaching purposes and is standard practice across the disciplines. Published literature is the lasting product of scientific research. It captures and documents the ideas, methods, results, implications and applications of projects and makes this information available to the broader research community and society to further research developments, grants, products, marketing, competitive advantage, *etc.* It is important for researchers to determine exactly when in their research process to disseminate their findings to the community and which of the many available avenues of communication is most appropriate. These decisions are influenced by place of work (academic, government, industry), job level, and practices in various chemistry sub-disciplines. The resulting published literature in chemistry is as varied and complex as the science it represents, and includes articles, patents, technical reports, conference proceedings, book chapters, and data sets.

Other complexities of publishing research lie in impact and prestige, discoverability and re-use, and availability and persistence. Tying one's name to research, being published and noted, is important to the success of many scientists. As purveyors of the literature publication process, publishers are also interested in procuring the most critical observations and ideas with the best potential. In addition to channeling the discovery of this research, they have high stakes in assuring the quality of research they publish and upholding the standards of scientific integrity. Peer-review is a long established and well-respected feature of scientific publication across most publishers. Clustering articles by disciplinary interest and novel potential further impacts discovery of worthy research. Well-respected publishers add value to the publication process through careful management of these and other editorial processes.

In addition to furthering knowledge itself, quality scientific research can also lead to new industrial applications and product development, improvements in scientific literacy and education, and informed public policy and national security. The field of chemistry is relatively unique, as it is both an academic discipline and an industry active in research and development. The extensive industrial sector is a heavy consumer of the published research literature, as well as a producer of its own research, primarily expressed in the form of patents. Commercial processes place special demands on presentation, authority, and accessibility of chemical information, which in turn significantly impacts the focus of government research and the experience of the academic chemistry research environment. In addition to publication of primary research, government contribution to the chemical information landscape includes high-quality data sets, standards for processes and safety, and education guidelines. Scientific societies such as the American Chemical Society in the US or the Royal Society of Chemistry in the UK play major roles in advocating and focusing on infrastructure for producing, re-using and building on quality scientific information.

The availability and persistence of published literature has a profound impact on the research process. Libraries and other information providers are concerned with the practical issues around discoverability and utility of published information. A variety of commercial and non-profit entities offer specialized tools to help researchers sift through the vast primary chemistry literature of journals, patents, registered compounds, and data sets. Abstracts are increasingly available online at no cost, publishers provide electronic alerts and news feeds, and conferences and social networks further highlight the availability of new research publications. In chemistry fields, most published content requires payment for access, reflecting both the expense to ensure quality and the potential for high-value re-use. With the advent of electronic information, pricing options have shifted from outright sale of copies to licensed access, which in turn has implications for ownership and responsibility of long-term archiving. Libraries remain major access points to and stewards of the chemistry literature; they maintain a high awareness of quality, and advise and collaborate with service providers.

In addition to providing researchers with access points to scientific information, libraries have historically taken on the task of preserving the scholarly literature to enable future use. It is easy to overlook the importance of older publications, but they constitute a significant portion of the accumulated scientific knowledge, and are responsible for supporting scientific development over the past several hundred years. In chemistry, where structural and reaction principles do not change drastically over time, older publications are very often still vital to current progress in a field, and in interdisciplinary research areas, past work is often re-considered from different perspectives. Research libraries worldwide store vast collections of journals in hard copy, often in state-of-the-art, climate-controlled, high-density storage facilities with sophisticated inventory control for easy retrieval. Publishers are also making digital back-files of older articles available for purchase or licensing, and libraries and publishers are working together to pursue preservation solutions, including the development of third-party archiving services, that will ensure access to the content in any future, foreseen or otherwise.

It is as important to develop good literature practices for your work as it is to improve your experimental and technical research skills. Good literature practices in scientific research require regular time spent reading or searching for journal articles and other relevant literature reviews. One should cultivate this practice to build competence in a new area, keep abreast of activity in areas of interest, become aware of exciting new possibilities and strong research groups, and scope out advantageous opportunities for collaboration and publication. Be aware of the scope of literature and information sources available to support both the theoretical and experimental developments of your research endeavors. The remaining chapters of this volume will introduce and guide you through a broad array of the most critical information resources and searching methods in chemistry research. It is well worth a systematic read to be aware of the landscape, and frequent referral for more focused guidance as you practice your research.

1.2 APPROACHING THE LITERATURE: PRINCIPLES TO BEAR IN MIND WHEN YOU ARE SEARCHING FOR CHEMICAL INFORMATION

Before proceeding farther into the landscape, there are a few general background areas worth delving into more deeply to better understand the literature resources you will use: basic information evaluation concepts; copyright and other intellectual property matters; how the published literature is structured; connectivity potential in the digital age; how libraries and other information providers can support your research; and the scientific input and approach you bring to your search process.

1.2.1 Scholarly Literature is Evaluated to Uphold Scientific Integrity and Vitality

A basic distinction of scholarly literature is that it has been evaluated to some extent before publication. It is important to the quality of one's own research process to ascertain up front the quality of related research in a discipline. The researcher must ultimately make the final determination if a work is worth looking at, starting with an assessment of how it has already been evaluated by the larger scientific community.

The most common type of primary publication of scientific information for academics is the journal article, and the first entity that decides what primary research is published in journals is usually the journal's editor-in-chief. Editors of scientific journals look for research that is original, scientifically important, and that fits the journal's scope in subject matter and treatment. Further review of manuscripts by published peers in the same research area serves to "flag what's important, set aside what's pedestrian, and abjure what's fraudulent".[1] A published article that has undergone a robust peer review and editorial process should contain data that tell a story and results that move the state of knowledge forward. The introduction of the article should set the stage for the story of the data analysis, and the novelty and intellectual interpretation of the research should be hammered home in the conclusion, giving a sense of the quality of thinking of the author.

Peer review is not a comprehensive evaluation system; reviewers do not generally repeat the experiments described, although review of supporting data is required in some characterization journals. The actual review process is not fail-safe and varies widely across publishers, which can significantly impact the reputation of a journal. The primary literature may be beset with a myriad of quality issues, including premature publication, lack of novelty, lack of focus or unclear explanation, inadequate review of the relevant literature, inadequate characterization of compounds created or altered in the research, missing or poorly designed experimental controls, failure to address alternate explanations, or unjustifiably strong statements.

Pre-reviewed research content is increasingly available online; conference proceedings, pre-print servers, research manuscript repositories associated with funding agencies, and community-supported, openly accessible and openly reviewed journals are a few of the examples. In the chemical disciplines, first disclosure and peer review of research findings carry significant weight in consideration of provenance, quality, and intellectual rights and are important considerations for the reputation and authority of the researchers themselves and particularly critical for commercial vitality in the industrial sector. Initial publication in an open or pre-peer-reviewed public venue may preclude later publication in journals with higher reputations or patenting to claim exploitable rights.

Even peer-reviewed journals vary widely in their reputation for quality and visibility of the research they publish, which in turn reflects on the reputation of the authors. One indicator of journal performance in contribution to scientific research is the number of citations by other research to the articles published in a particular journal. This principle underlies the Thomson Reuters Journal Impact Factor, which is often used by a broad range of literature users such as publishers trying to attract authors, institutions considering tenure for research faculty, researchers identifying top journals to monitor, and libraries attempting to prioritize access and preservation of journal content. Discovery service providers also consider the provenance of published literature and data, but tend to include a fairly broad approach to sources to give the chemical researcher the fullest information of the activity potential in their research area. Promising new journals may not be indexed until they have proven their potential, maybe through a high Journal Impact Factor, which takes two years to calculate.

1.2.2   Data Provenance and Evaluation is a Critical Part of the Research Process

Many research areas in chemistry generate and analyze significant volumes of data. Data associated with chemical research can appear directly in articles, in supplementary files referenced by articles, as part of compiled data sets, and in repositories of specialized types of chemical information. The provenance and quality of compound characterization and other published data are particularly important to chemistry research. Results and interpretation are only as good as the data on which they are based, and their potential for meaningful contribution to scientific knowledge depends on their correlation to other evidence or revelation of abnormal observations. As you work with both your own data and those you are re-using from other sources, it is critical to ascertain that they actually represent what they are purporting to and are reliable, based on the quality of the measurement process. The opportunity to apply promising methodologies on large production scales in the commercial sector hinges on adherence to standards and regulations of practice. You can imagine areas of chemistry, such as the development of drug formulations and construction materials, where lack of attention to safety, consistency, and reliability can not only compromise the outcome of the experiment but could potentially endanger vast numbers of people.

Quality data start with robust data collection practices, including documentation, using multiple sources of measurement, calibration of equipment, and using controls and/or standard reference data. It is most important for users of data to know how it was collected to determine if it is relevant, if it actually measures what was intended, and if its collection was executed in a sufficiently accurate and precise manner for re-use in the new context. Good documentation should include careful notation of all the parameters in which the data were measured, including equipment, conditions, methodology, characterized standards, and experimental context. Multiple sources of a measurement re-enforce the quality of the measurement technique and specific execution, and normalize inherent variability within and across chemical systems. Calibration to well-characterized standards also maximizes the technical quality of a measurement. The use of controls within an experiment or comparison of results to standard reference data establishes the value of the measurement that is distinct to a sample and of interest for further analysis. For example, the use of standard reference data to identify values related to specific structural characteristics of compounds is relevant to spectra searching, for example.

The National Institute of Standards and Technology (NIST) concerns itself with supporting robust chemical and physical data evaluation and addresses standards across four stages: data collection, basic evaluation, relational analysis, and modeling.[2] How data is collected, documented and stored can impact later accessibility to that data. Basic evaluation questions generally focus on the reproducibility of the data using the same collection methods. Relational analysis is concerned with consistency of the data at hand with other data that describe the material, such as related properties or independent reports of a particular property. Modeling calculations can indicate the predictability of the data as an indicator for this property under the conditions at hand. In practice, processes for assessing and assuring quality of data are especially well developed in materials research and production. Depending on your need when looking at published data, you might require quality indicators ranging from general specifications for a class of material to certified standards of specific compounds. In active research, you might find yourself working with commercial data with specifications provided by the manufacturer, or with preliminary data from collaborating projects.

NIST provides a decision tree to classify property data and determine appropriateness in the context of purpose and use. This protocol is freely available as a simple interactive assessment tool originally developed for the NIST Ceramic WebBook and is a reasonable check-list when working with any published data where quality and provenance is a consideration.[3] Indicative questions for literature and data evaluation include:

- Is the source journal peer reviewed?
- Are the experimental methods adequately described to be repeatable?
- Are any compounds characterized well enough to identify?
- Are the results consistent with other indications in the published literature?
- Does the explanation build on previously published research?
- Do the authors address alternate explanations of the data with further experiments?

As with the scientific research process in general, the provenance of the resulting observations and explanations is important when considering whether the information is of sufficient quality. If little is known concerning the who, what, why, where, when, and how aspects of a research project, it could be considered of indeterminate quality and therefore unacceptable for reference. Referencing the original source of the data, as well as any available provenance, lets the reader make a judgment about the quality and applicability of these data.

Data management is of increasing interest to research-granting agencies, including the National Science Foundation (NSF), which as of 2011 requires all granted projects to include a data management plan. In 2009, an Interagency Working Group on Digital Data developed recommendations for managing data, including some general components to consider for a management plan: "provide for the full digital data life cycle and...describe, as applicable, the types of digital data to be produced; the standards to be used; provisions and conditions for access; requirements for protection of appropriate privacy, confidentiality, security, or intellectual property rights; and provisions for long-term preservation".[4] More or less specific guidelines are being developed by the various US funding agencies; the NSF is primarily leaving this to be determined at the level of peer-review and program management to reflect best practices for disciplines and other "communities of interest".[5] The provenance documentation practices discussed above should be rigorous enough to cover most data management plan requirements.

1.2.3 Scientific Literature is Considered Intellectual Property

Ultimately, the purpose of scientific research is to contribute to the greater scientific knowledge base in a useful way and lead to applications for society. The ideas and efforts towards this process are considered property of an intellectual nature and are governed through their documentation. The legal framework of intellectual property is to translate the association of scientists with novel ideas and processes into terms that can serve in the practicable everyday world of business, including documentation for provenance and remuneration. In legal terms, intellectual property is about ownership and the potential benefits therein. It was designed by Congress to address Article 1 of the United States Constitution: "to promote the Progress of Science and useful Arts, by securing for limited Times to Authors and Inventors the exclusive Right to their respective Writings and Discoveries".[6]

Novelty is a core consideration in supporting scientists' and companies' rights to own an idea or a process. The definition of novelty in most jurisdictions is delineated by first public disclosure: anywhere, in any venue, for any purpose. Because of the high potential for value, most publishers in the field of chemistry will not accept work that has been extensively disclosed in a public venue. Patent applicability can hinge on the date and nature of disclosure and becomes especially critical when coordinating rights globally. Ideally, the first public appearance of an idea that is well enough researched to enter the scientific record should be well documented, most often in a published article or patent application. These forms of communication are readily citable, with fairly rigorous presentation of content. However, the first public disclosure of one's research may often be much less rigorous, such as a presentation at a conference. As a result, chemists need to be mindful of future plans to publish in journals or file patent applications as they prepare their presentations.

Scientific research, particularly chemical research, is expensive. Public and private monies earmarked for basic research are available competitively. The chemical industry is interested in productive chemical technologies to make a return on the investment of development. Publications, including patents, are professional scientists' and chemical companies' key to sustainable funding and growth through claim to ownership. Most scientific publications are considered under one of two flavors of intellectual property, copyright, or patenting.

1.2.3.1 Copyright. In its legal form, copyright is at least two levels removed from the everyday world of scientific research. It does not relate to experimental design, nor does it contribute to the process of good writing. For most authors, it only seems to come into play when one is trying to publish, and then it often appears as a barrier. Why would a chemist want to have anything to do with copyright or even think about it? It comes down to basic issues surrounding the sharing of creative work with others and, in turn, re-using their work. Your greatness as a scientist lies in your ideas, but these remain in your head and might as well be mist unless you express them in a form that resonates with those whose attention you want. Once your audience takes notice, it will be of the idea, and, in the excitement, you want to be remembered as its originator. Copyright law provides a recognition stamp for a piece of work that captures an idea and governs the ways in which these ideas may be re-used by other scientists.

Copyright protects the expression of any creative act such as music, art, journalism, fiction writing, and many other endeavors where people may want to seek compensation and/or credit for their work. The author originally owns the rights to his or her work, meaning that, for the work to be "copyrighted", he or she does not need to do anything more

formal than capture it in a tangible medium (including online). However, as a legal tool, copyright must be able to stand up in court if the
rights of ownership are in dispute. Every researcher hopes their work
will be of sufficient interest in his or her discipline that it will be discovered and read by other researchers, granting agencies, and chemical
businesses. The potential value of a paper is tied up in where it is
exposed and what can then be done with the content, activities overseen
by copyright. As the initial copyright owner, the author needs to consider how best to manage the exposure and re-use of the work to meet
his or her personal and professional needs.

Copyright is automatically assigned to an idea "the moment it is
created and fixed in a tangible form that it is perceptible either directly
or with the aid of a machine or device";[7] the rights and opportunities
thereby granted are up to the owner to manage and stipulate to the
public world. Currently, one of the primary roles of scientific publishers
is to formally establish the first public disclosure of a work that invokes
those rights, and reputable publishing houses are knowledgeable in both
the scientific discipline and the ways of copyright. Publishers also provide additional value by coordinating with the vast network of publishing peers in a discipline to review the quality of the contribution and
by placing the work among others of good quality in reputable journals,
thus increasing the collective potential to be noticed by the right people.
To manage and guarantee all of these services, publishers want a specified relationship with copyright that oversees the legal status of all
these activities. In exchange for publishing your article, most scientific
publishers will require transfer of your copyright: in effect, transfer of
ownership of the work. As the original copyright owner, you always
have the option to self-publish if you are prepared to manage your
rights, the evidence of first disclosure and any further development and
if you believe your work is strong enough to stand on its own.

For the vast majority of scientific articles published in traditional
journals, once a manuscript is accepted for publication, it is likely that
the authors will be asked to sign an agreement or contract that includes
language regarding the copyright of the work. Many contracts require
the author to transfer copyright to the publisher, meaning that they will
then own all the rights to the article. To do anything further with the
article, authors and readers alike will need to seek permission from
the publisher as the new rights holder. This includes posting copies of
the article on a website, sharing it with colleagues, and using figures in
presentations or classes, even if the author is the one teaching them. It
also includes reusing any of the content subsequently in a thesis or
dissertation. Given the original intention of copyright to support the

creativity of the original author and the rather dire impact of cutting you off from your work by transferring all such rights, many publishers will return several rights under the same contract, generally giving permission for the author to share copies with individual colleagues and re-use figures in presentations, classes and dissertations. Because the publisher continues to be the copyright owner, they will usually ask you to provide a citation or a copyright notice in the new venue for any part of your article that you re-use. The American Chemical Society presents FAQs and other learning materials on copyright for publishing authors.[8]

It is always an option to seek permission to do anything that is not specified in a contract, and most scientific publishers will grant this for non-profit oriented uses, especially by the original authors. To use other people's work, you will also need to seek permission from the copyright owner. It is not usually difficult to gain permission for common types of re-use, such as reproducing figures or quoting a brief section of text, many publishers now have automatic permissions systems, such as the RightsLink service used by the Publications Division of the American Chemical Society (http://pubs.acs.org/page/copyright/permissions.html) and other major publishers, which can be used to grant permission for certain pre-determined uses. It is important to note that the requirements for re-use will differ from publisher to publisher, so it is important to follow the form through to the end. Individual scientists in academic institutions making copies of articles (print or digital) for their own general reading purposes usually do not need to seek direct permission from copyright owners to keep these copies. This type of use is provisioned in the Copyright Act as "fair use". The Fair Use provision addresses a number of types of re-use commonly associated with academic, educational and other non-profit endeavors, such as limited and restricted copies for individual research and teaching. The general understanding is that the use will be small scale and not translate to commercial potential that is still protected for the owner. For more information on acceptable fair use, see The Factsheet on Fair Use,[9] the Circular 21 from the U.S. Copyright Office,[10] or consult a legal authority.

1.2.3.2 Managing Rights in the Digital Environment. Rights associated with intellectual property are not defined relative to format or genre. However, in the digital environment, the scope of the playing field is changed. There is much broader access potential and a much richer technical environment for re-use and re-purposing of content, such as in data-driven research. Simultaneously, the global political and economic environment has encouraged increased participation in scientific research and the chemical enterprise. There are vastly more scientific manuscripts produced than the expanding journal options can absorb, and the peer-review system is swamped. There is a rapidly increasing readership and increasing pressure to publish manuscripts directly online to increase speed and availability. Emerging data-driven approaches to research and development demand greater technical treatment and access to content.

Players on the field have responded to these drivers accordingly by intensifying their approaches with overall compounding effects on the flow of information. Higher potential for global-reaching commercial value coupled with perceived higher competitive threat spurs content owners to tighten rights management measures. In the absence of acceptable standard practice, such measures have tapped into other legal tools such as contract law, and technically based restrictions on access and use, currently enforced through the Digital Millennium Copyright Act (DCMA). Typically, these restrictions limit use far more than with analog information sources. The most visible restriction to researchers is the amount that can be downloaded from various information sources, including database result sets, journal articles, and book chapters. Printing, saving, filing in reference management tools, or forwarding to colleagues may all be restricted or disallowed altogether.

There are other subtler, but no less critical impacts on long-term access and use as specifications of ownership and hosting of the scholarly literature are shifting. Most electronic scholarly journal content is made available to users through license rather than sale as print subscriptions had been. Libraries have negotiated new terms for access in perpetuity to fulfill their mission to make sure that articles are available in the long term. Since publishers remain the content owners, they, rather than libraries, are now also responsible for archiving. Third-party services are emerging to support the ongoing technical integrity of electronic information.

The online environment has increased the potential for the sharing of work; however, it is still important to the integrity of a work to manage the rights of re-use and provenance even if the content is openly available for the initial use of reading. Creative Commons is a non-profit organization developing a new approach to managing and communicating terms of copyright of work in the digital space. The underlying principle is that the work will be openly available for public dissemination and use with a variety of conditions specified by the owners. Several licenses are available with various combinations of specifications for attribution, sharing and commercial purposes. Creative Commons licensing is based on copyright and provides the legal code to uphold it.

Additionally the licenses include versions of the terms expressed for owners and users not legally trained and also in machine-readable form to communicate and functionally enable rights and permissions in the digital context; see http://creativecommons.org/licenses/ for more information. As the global legal climate surrounding intellectual property establishes itself in the digital environment, content authors, owners, and users juggle a complicated information landscape.

1.2.3.3 Ethics. Authors have certain ethical obligations to the scientific enterprise. Publishing contracts will often include requirements that the work submitted presents original research, an accurate account of the research performed, and an objective discussion of its significance. They further stipulate that all coauthors must be aware of the submission, that the authors submit their work to only one journal at a time, and that they disclose the submission history of the manuscript.[11] Original work should not plagiarize text or figures from other published works, even if prepared by the same authors. The tendency towards self-plagiarism is particularly problematic as researchers build on their own previous work, but each newly published work should have enough novelty to stand as a separate and distinct contribution. Connections to previous work, by the authors or others, should be fully attributed and referenced. Permissions for more extensive use of previous content, such as figures in a review article should be sought from the copyright owner, as discussed above. Such practices constitute a code of conduct and personal responsibility that is core to the definition and ongoing integrity of chemistry research. For further reading on best practices for scientists, see "On Being a Scientist", freely available from the U.S. National Academy of Sciences.[12]

1.2.3.4 Patenting. Patenting is another approach to intellectual property that focuses on the design of technology, human-invented approaches to accomplishing a specified task. This type of intellectual protection involves a different form of documentation, and the resulting patent literature constitutes the primary contribution of the chemical industry. Rights owners are trading public disclosure of their approach for a limited period of exclusivity to develop any commercial potential. Patents allow the public to benefit in the longer term through healthy competition and additional development, while still supporting the pursuit of commercial viability by the originator. Otherwise, owners of commercial processes might keep successful technologies secret indefinitely. A granted patent supports this right for the first party to file, even if others come up with similar ideas independently, as long as the invention is novel. The United States also requires that the invention have utility and offer a non-obvious change to existing technology. Assignees have twenty years to develop and market the technology without competition should they pursue it.

The chemical syntheses and refinement processes developed in industry are patentable, which makes the window of exclusivity a highly valuable right in the commercial sector. As a result, patents are carefully construed to cover a broad a range of potential approaches within each technology to give companies flexibility and multiple stepping-stones to pursue. Technologies developed within the scope of academic research are also patentable, and universities will often contract with commercial partners to scale and market promising technologies. A few technologies out of millions of patents prove to be of high market value, and the owning companies will fiercely defend their exclusive advantage. While development rights are exclusive, the disclosed design is public information, and, although the patent is written in such a way as to obfuscate the critical pieces as much as possible, it can still be very useful for indicating the direction of proprietary research in a given area, as well as providing other important chemical information, such as characterization properties. As a result, patents are a rich body of chemical literature publically available to every research chemist and worthy of serious consideration; approaches to using patent literature are more fully discussed in a later chapter of this book. For further reading on patenting relevant to chemistry, see the handbook "What Every Chemist Should Know About Patents", available from the American Chemical Society.[13]

1.2.4   Scholarly Literature is Structured to Facilitate Research

1.2.4.1   Primary Literature.   The first time an observation or idea appears in a public medium constitutes first disclosure and is categorized as primary literature. This is the important point for discovery and the critical point at which an idea has enough scientific potential behind it to become part of the development of a scientific discipline: "if your research does not generate papers, it might just as well not have been done".[14] The primary literature represents the state of a research area and will supply you with information on methods and protocols. In chemistry, many primary publications appear in the form of research articles, clustered in journals ranging from general or multidisciplinary to specialized by sub-discipline, methodology, or nationality. Patents, conference papers, and technical reports also constitute a significant portion of the primary literature globally across the chemistry sub-disciplines. The authors, editors, and reviewers of the various primary resources have reviewed the information and deemed it publishable, but it remains to the researcher to locate it and decide if it is relevant to his or her own work.

1.2.4.2 Secondary Literature.   Over one million primary publications are indexed by the Chemical Abstracts Service each year in chemistry and its related fields.[15] It is not possible to follow the developments or even find relevant information in any one area without additional organizational tools. Publications that parse, abstract, index, or otherwise break down and group the information and ideas appearing in the primary literature are categorized as secondary literature. There are two general types of secondary literature, depending on the content and purpose. Abstracting and indexing services facilitate research of ideas by organizing the bibliographic information of the primary literature. These tools tend to be large-scale resources, covering a broad range of primary sources to facilitate multidisciplinary and comprehensive research. Databases extract and aggregate specific information from the primary literature to create high-value collections of experimental, analytical, or preparative information. These collections tend to be fairly specialized by type of information or research methodology.

Opportunities for searching in an area of interest simultaneously across multiple information sources and types are becoming more prominent in the web-enabled, digital information environment. Chemical Abstracts Service is one of the most prominent secondary literature providers, specializing in thorough coverage and indexing of the chemistry literature through a variety of systems, including SciFinder and STN (Science & Technology Network). SciFinder links different types of bibliographic, characterization, and preparative information from within the primary literature to enhance the research process from idea to experimental design. Successful use of the secondary literature tools will contribute to your knowledge of a research area. Developers of these tools carefully manage the inclusion and organization of primary literature sources based on scope and perceived quality, but no additional value-based judgment is offered beyond this. The intellectual process of identifying what specific articles and information is relevant information remains to the researcher.

1.2.4.3 Tertiary Literature.   Even with the vast number of primary publications in the chemistry-related disciplines and the wide variety of secondary tools available to navigate them, a scientist may still seek additional input to ascertain the gestalt of the research in an area before trying to search it directly. Such scenarios could include a scientist pushing into an unfamiliar research area, a lab group changing its approach to an experimental methodology, or a chemistry graduate student learning to practice research. There are several types of literature in chemistry designed to give an overview of a research area, methodology or practice, these resources are referred to as tertiary literature. Review articles and chapter-books give an overview of a research field at a given time. They are written by experts in the field, long-time practicing scientists, and can cover the development of the primary theories, branches into other fields, applications in industry, primary educational models, future directions with high research potential, and even research lines that didn't work out. Treatises and handbooks meticulously review the developments of specific research methodologies or experimental best practices in various areas of chemistry, such as organic synthesis. Graduate-level texts, encyclopedias and other primers, such as this book, are another type of tertiary literature designed to introduce an inexperienced researcher to a particular field. Tertiary literature sources offer expert value-based judgments of the published literature and assessment of data in the research area under consideration. It is important to keep in mind that these sources are out of date as soon as they are written in terms of the state of the science in any given area; they are a great starting point to a new area of research but not a robust finishing point for preparing your own experiments and publications.

1.2.5 The Literature is a Web of Potential

Each published article has potential in the scientific enterprise, waiting to be found and read by another scientist who sees its potential and can build on it. A key aspect of this path to successful contribution is how other scientists who would be interested in the content of an article happen upon it. An early part of the discovery process for many researchers is the groupings of articles that make up issues of journals that are read regularly. There are many other points of connectivity; the units of the primary literature and the research experiments, observations and conclusions that they represent do not exist in isolation within their host journals. Research articles and patents build on previous reportings, and, in turn, influence those who subsequently read them; the scientific ideas in each article are linked to other published articles. There are many different ways that individual scientists approach their literature practice and process of finding new articles of relevance to their current research projects. However, they are all based on some kind of link from one article to another, one scientist to another, or one idea to another, with each subsequent link related to the former in some way.

For a specific research project, an idea may start with one article read by a scientist. The scientist may then read some of the article's references for better background, then find papers that cite the starting article to see how others have built on it, then examine articles that cite the same references as the original article to see how others have built upon the earlier research, and so on. Much like a pearl that builds up in layers upon the initial stimulation of a grain of sand, this technique of building up a cadre of articles and research awareness through following links is referred to as "pearl growing", or "the Iterative Approach to literature searching."[16] Common link paths highlighted by the discovery services in the secondary literature include journals, publishers, authors, institutions, sub-discipline, methodology, type of application, compounds, and physical properties, as well as both references and citations. It is the prerogative of the researcher to navigate the various paths to find the best literature for their particular purpose. The networked online environment is having a profound impact on the ability of researchers to move along these links to aid discovery of information and build knowledge bases. The majority of chemical information resources are available online. As more standards emerge and develop for encoding text and other information to appear on the Web, more links are being activated between common information elements across resources that go well beyond the traditional journal, author, and references.

Chemical information is in a unique position in terms of development potential in the online environment, influenced by a variety of factors that complicate the realization of this potential. The chemistry field is actually one of the earlier pioneers of online representation of information, with machine-readable encoding systems for chemical compounds dating back to the line notation systems of the late 1940s. Chemical information is also exceedingly complex and nuanced in what it represents; structural characterization of compounds, chemical and physical properties of compounds, preparation and purification methodologies, and analytical techniques are all considered by chemical scientists in their research. This intensity around information has been accompanied by elaborate representation schema for various aspects of the information since the heyday of alchemy. In 1919, the International Union of Pure and Applied Chemistry was formed to more systematically consider and review chemical information representation and apply standardization in some critical areas internationally, including chemical compound notation for both human and machine reading purposes.[17] The latest example of efforts in this area is the IUPAC International Chemical Identifier (InChI), which provides interoperable chemical structure encoding between different publishers and chemical information systems.[18]

Robust and standardized machine-readable encoding of information has also enabled the emergence of new and powerful data-driven approaches to research. Informatics, as this type of science is generally called, is touching on many fields, including chemistry. Research processes that were previously managed by the researcher, such as data collection and management, are increasingly automated, and ultimately the computer can activate a variety of links among and between data sets to indicate patterns of potential interest. It is still up to the human researcher to make some determination of the value and to pursue further research of any of these patterns.

As these computer systems become increasingly sophisticated, they are beginning to perform more of the valuation themselves, "learning" from patterns of previously assigned values and performing self-assessment based on error rate analysis. This area, in which the computer applies value-based analysis to research input, is referred to as semantic processing. This approach is not only being applied to numeric or other non-textual research data, but to the linking patterns used by scientists when searching the literature, as well as the early stages of analysis of text in the primary literature and, by extension, a kind of analysis of the intellectual contribution of individual scientists. This sounds very much like the literature research process for individual humans that we have been discussing throughout this chapter. What could be lost with the automation of more processes formerly performed by educated chemists, and what more could those researchers do beyond what is possible now with more time freed from automated tasks? As more data, including the direct intellectual contribution of researchers, is presented online and linked to other information, pattern recognition and evaluation is enabled and the impact of these considerations will become increasing prominent. There certainly are implications regarding productivity value and re-use of material considered to be intellectual property and therefore protected by copyright or patent law. There may also be implications for what is considered by the chemistry community to be acceptable standards of practice when balancing machine and human analysis and valuation to further the research enterprise.

1.2.6   Libraries and Other Information Providers Offer Disambiguation

Amidst the complexities and complications of the chemical information landscape, libraries focus primarily on enabling use of scholarly materials. An ideal goal for searching the literature for researchers and information providers to strive for might be 90% unassisted use 24/7 anywhere, complemented by detailed support the remaining 10% of the time. Information providers are in the business to consider highly dis-intermediated experiences for researchers to enable the most efficient approach within a researcher's individual process and point of need. Both content and access are key components of a dis-intermediated research process, through combination of clearly defined scope of content, expert curation, value added content analysis, and automated organizational structure. Expert curation is the highest value added to most chemistry resources, involving scientists and other field experts to determine what content to include and highlight, what links to include and highlight and how to put these together to clarify the opportunities and potential indicators for researchers.

Researchers' needs not covered by 90% solutions require expert assistance. These needs should not be underestimated; they could translate to "aha moments" for researchers, critical learning opportunities for students, or indicators of emerging areas of chemistry research and potential in the information landscape. The questions you are asking may be cutting edge and unique enough to not be represented in standard ways in searching tools. In a well-meant effort to maximize the opportunities of the online environment, database and information providers often try to make tools more intuitive. In reality, expert search functions are often diminished, resulting in more difficulty finding relevant information. If you have spent over 20 minutes in fruitless searching, this is not good use of your time; ask for help. There are experts who search for information for a living; they often have access to better tools and have invested time to develop better work-arounds; they can save you a lot of time.

This volume is authored by chemistry-focused librarians across the United States and Canada who perceive a need to more broadly support graduate students and researchers in chemistry with their literature use. In addition to expertise in the literature landscape of chemistry, librarians have access to networks of other experts, and participate in a variety of services and activities to further broaden both the support and expertise they can provide. They curate specialized finding tools in chemistry, such as properties finders and virtual shelf browsers; offer training, guides, and feedback opportunities with specific resources and search techniques; and actively participate in scientific societies and liaise with publishers and other professional development programs for chemists. All of this expertise is only as good as it is useful for chemists; we welcome the opportunity to assist your literature research in a variety of ways. Another useful volume addressing the broad issues of publication is the ACS Style Guide, 3rd edition published in 2006 by the American Chemical Society.[19]

The balance of supporting researchers in a robust searching process through independent options coupled with specified assistance represents a moving target as the research landscape continuously changes. Iterative development is critical for information providers to aim for a successful highly dis-intermediated environment. Follow-up analysis of assisted experiences is needed to assess what is indicated about gaps in dis-intermediated solutions or potential new service areas. Such are the requirements of robust information systems and services and chemistry information providers tend to invest significant resources into ensuring robust content, organization, support, and other added value. As the digital markup of chemical information improves, more direct engagement is possible with non-tactile literature and libraries transition support of print-based research processes to online-based research processes.

1.3   GETTING STARTED WITH THE CHEMICAL LITERATURE

1.3.1   Your Literature Research Is Only as Good as Your Input and Process

A literature search is a significant part of the overall research process. It is up to you to leverage the structure of the literature, discovery tools, pearl growing, valuation, and good tracking skills to tap its potential. If you do not take the time and care to plan your process up front, you will quickly be swamped by the vastness of the literature, and likely miss key findings or painstakingly recreate experimental methods previously published. Please remember Frank Westheimer's aphorism, "Why spend a day in the library when you can learn the same thing by working in the laboratory for a month?"[20]

When searching through the literature, the information you have in hand – previous research, active authors, chemical structural information – can serve as starting and linking points. Since your search of the literature may be for background information, a comprehensive sweep of previously characterized compounds of interest, a specific set of physical properties, or a particular synthesis route, what you already know will help identify which information resources are best suited to help. The remainder of this book provides some description of the more commonly used chemical information resources designed to help the researcher determine which to use and how best to get started for various needs.

Given the complex nature of chemical compound characterization and the breadth of research fields that touch on chemistry, some types of chemical information are more complicated and require advanced searching methodologies. Good starting places and best practices for more specialized searching are detailed in the later chapters of this book. This is not a comprehensive sweep of all potential approaches to searching in chemistry, so as you specialize in your area of research, becoming thoroughly competent in the relevant advanced searching methodologies will be critical for a robust research program.

Reviewing and assessing the results requires an understanding of what additional relevant information may be available, evaluating new search leads, such as other associated compounds, and recognizing better index terms. Reviewing specific result records will indicate what can be expected in that information resource, and gives a sense of how structural, reaction or property information is encoded. To quote from the conclusion of the physical properties chapter: "important skills for a searcher are persistence, creativity, and a sense of what avenues are most likely to be successful and which ones are unproductive... not unlike the qualities of a good detective".[21]

1.3.2 How to Use the Literature to be a More Efficient Chemist

So what are some practical tips for mastering your work with the chemistry literature? At Cornell University, we have created a guide titled, "7 Ways to Be a More Efficient Chemist" that boils down several key activities you can set up right away to help yourself in the literature aspects of your research (http://guides.library.cornell.edu/7chemistry, original guide by Kirsten Hensley, 2008). The guide points to specific resources at Cornell University, but the principles apply anywhere for any chemist at any stage of research.

1.3.2.1 Streamline Your Connections to the Literature Resources You Use Regularly So You Can Access Them Anywhere, Any Time, and from Any Device. Most research libraries have a proxy system in place for connecting to resources when you are off-campus; many also provide bookmarklets or apps for re-loading web pages with your institutional authentication so you can log in from anywhere. Set up bookmarks in your web browser of choice, or use a webroot or some other system with your most frequently and regularly used resources, using the links provided by your library, which should include the proxy authentication. Apps covering a variety of literature resources and searching options are also increasingly available if working on smaller mobile devices fits into your work style.

1.3.2.2 Organize the Hundreds of Articles and References You Collect in Your Literature Research. Many citation management programs are available with various organizational features and costs ranging from free to reasonable educational discounts. You can group references by topic, project or specific question you are researching. Most will import PDF files and some will pull out the bibliographic information for you so you can organize the papers. Some allow for collaborative work. Most literature databases will export references in formats directly importable to these programs; some programs can even be used to search other content or linked into directly.

1.3.2.3 Regularly Monitor the Contents of the Top Journals in Chemistry and Your Specific Sub-discipline Once You Start Actively Researching. Most scientific journals provide email or RSS feed alerts of issue content for free. JournalTOCs (http://www.journaltocs. ac.uk/) collects thousands of feed links to scholarly journal tables of contents, and you can create groups of journals to monitor from this free service. If you are not familiar with the journals in a particular sub-discipline, you can get an initial list to start by exploring the Journal Citation Reports ISI Impact Factor rankings if your institution subscribes to this assessment tool. These rankings are based on numbers of citations to a journal relative to the number of articles published within a fixed time-frame, roughly indicating how much impact the research published therein is having on informing further research in a given area. Review journals tend to show the highest impact with this measure, as they are broad in scope and can be particularly helpful for reference when new to a research area.

1.3.2.4 Set up Alerts in the Literature Databases to Monitor New Research by Topic. This technique will cut across journals and other literature sources and allow you to zero in on specific methodologies or compounds of interest on a more specific level. Most databases, such as SciFinder, Web of Science, MEDLINE, *etc.*, offer alerts based on your searches of interest. You can also save searches and come back to them to build up a critical mass of literature in an area to export to your citation management program.

1.3.2.5 Read Books and Review Articles for Background Material. You will be expected to build up knowledge of various areas pretty quickly as you begin more research. These could be the state of current research areas, chemical reaction or other experimental methodologies, or potential for application. Treatises and review journals as mentioned above are available that cover all these types of information, as well as periodic review articles in primary journals for more specific or timely topics.

1.3.2.6 Be Familiar with the Options for Acquiring the Full Text of Articles through Your Library or Information Center. Most research libraries have fairly robust collections of electronic journals that will be directly available to you or will provide document delivery for needed articles. Finding these links among thousands of others will vary by local institution. No research library has direct access to all published literature, digital or hard copy, but there are a number of collaborative systems that research libraries use to make content available among institutions. Most libraries participate in some kind of interlibrary loaning system for hard copy, photocopies, and increasingly for electronic content as well. Systems for article sharing tend to be national or international, many regional approaches also exist for books, including service from joint storage facilities.

1.3.2.7 Ask for Help from Librarians with All of the above Tasks and More. If we don't know specifically how, we will find the right assistance for you. This is the top priority and core responsibility of the public services librarians in any library. Most research libraries will have librarians who specialize their service in key disciplines, including chemistry, which tends to be a literature-heavy discipline.

1.3.2.8 Bonus: Be Aware of Specialized Electronic Reference Resources for Reaction Specifications, Physical Properties, and other Scientific Data. More and more of the data supporting chemistry research are becoming available in online venues. The traditional reference collections in research libraries supporting chemistry tend to be expansive and well used but cumbersome and probably not as well discovered as they could be for supporting experimental and technical work. As these resources become more available online and libraries are able to support them, it can have a positive impact on your workflow.

Overall, remember that the library is intended to support your literature research, in accessing content, improving your searches, and helping you become a more efficient and better prepared chemist.

REFERENCES

1. J. Polanyi, Why our discoveries need to surprise us, *Globe & Mail*, Oct 3, 2011, pA17. [Cited by *ACS On Campus*, New Mexico, Jan. 13, 2012.] http://acsoncampus.acs.org/past-events/#unm

2. R. G. Munro, *Data Evaluation Theory and Practice for Materials Properties*, U.S. Dept. of Commerce, Technology Administration, National Institute of Standards and Technology, [Gaithersburg, Md.], 2003. http://purl.fdlp.gov/GPO/gpo3836. Accessed 31 October 2012.

3. http://www.ceramics.nist.gov/IDELA/IDELA.htm. Accessed 31 October 2012.

4. *Harnessing the Power of Digital Data for Science and Society*; Report of the Interagency Working Group on Digital Data to the Committee on Science of the National Science and Technology Council; 2009. http://www.nitrd.gov/about/harnessing_power_web.pdf. Accessed 31 October 2012.

5. *Data Management & Sharing Frequently Asked Questions*; National Science Foundation; updated November 30, 2010. http://www.nsf.gov/bfa/dias/policy/dmpfaqs.jsp. Accessed 31 October 2012.

6. *United States Constitution*; Article 1, Section 8. http://www.archives.gov/exhibits/charters/constitution_transcript.html. Accessed 31 October 2012.

7. *Frequently Asked Questions about Copyright*; U.S. Copyright Office; revised July 2012. http://www.copyright.gov/help/faq/. Accessed 31 October 2012.

8. American Chemical Society, Joint Board/Council Committee on Publications Subcommittee on Copyright, *Frequently Asked Questions about Copyright*, revised August 2010. http://pubs.acs.org/page/copyright/learning_module/module.html. Accessed 31 October 2012. *ACS Style Guide: Effective Communication of Scientific Information*; Coghill. A. M., Garson, L. R., Eds.; American Chemical Society: Washington, DC, 2006.

9. *Fair Use*; Factsheet FL-102; U.S. Copyright Office, Library of Congress: Washington DC, reviewed June 2012. http://www.copyright.gov/fls/fl102.html. Accessed 31 October 2012.

10. *Reproduction of Copyrighted Works by Educators and Librarians*; Circular 21; U.S. Copyright Office, The Library of Congress: Washington DC, revised November 2009. http://www.copyright.gov/circs/circ21.pdf. Accessed 31 October 2012.

11. American Chemical Society, Publications Division. *Ethical Obligations to Publication of Chemical Research*, revised June 2102; Washington, DC, 2012. http://pubs.acs.org/userimages/ContentEditor/1218054468605/ethics.pdf. Accessed 31 October 2012.

12. *On Being a Scientist: A Guide to Responsible Conduct in Research; Committee on Science, Engineering and Public Policy*; National Academy of Sciences, National Academy of Engineering, and

Institute of Medicine of the National Academies; National Academies Press: Washington, D.C., 2009. [Cited by *ACS On Campus*, Cornell University, Apr. 20, 2012. http://acsoncampus.acs.org/past-events/#cu] http://www.nap.edu/catalog.php?record_id=12192. Accessed 31 October 2012. Macrina, F. L. *Scientific Integrity: Text and cases in responsible conduct of research*, 3rd edition. ASM Press: Washington, DC, 2005.

13. American Chemical Society, Joint Board/Council Committee on Patents and Related Matters, *What Every Chemist Should Know About Patents*, revised 2002; Washington DC, 2002. http://portal. acs.org/portal/PublicWebSite/about/governance/committees/WPCP_006903. Accessed 31 October 2012.

14. G. Whitesides, Whitesides' Group: Writing a Paper, *Adv. Mater*, 2004, **16**, 1375–1377. http://onlinelibrary.wiley.com/doi/10.1002/adma.200400767/abstract. Accessed 31 October 2012. [Cited by ACS On Campus, Cornell University, Apr. 20, 2012. http://acsoncampus.acs.org/past-events/#cu]

15. American Chemical Society, Chemical Abstracts Service, *CAS Statistical Summary 1907–2007*; Columbus OH, 2008.

16. C. Huber, "Lecture 2: Techniques of the Efficient Information Searcher," From: *Chemistry 184/284: Chemical Literature*. University of California Santa Barbara. http://legacy.library.ucsb.edu/classes/chem184/184leca2.html. Accessed 31 October 2012.

17. International Union of Pure and Applied Chemistry. *About IUPAC*. http://old.iupac.org/general/about.html. Accessed 31 October 2012.

18. International Union of Pure and Applied Chemistry. *The IUPAC International Chemical Identifier (InChI)*. http://www.iupac.org/home/publications/e-resources/inchi.html. Accessed 31 October 2012.

19. *ACS Style Guide: Effective Communication of Scientific Information*; Coghill. A. M., Garson, L. R., Eds.; American Chemical Society: Washington, DC, 2006.

20. F. H. Westheimer, Harvard University, Cambridge, MA. Quoted in a press release, "Frank H. Westheimer, 95, Pioneering Harvard Chemist, Dies", 2007. http://www.fas.harvard.edu/home/news-and-notices/news/press-releases/release-archive/releases-2007/westheimer-04162007.shtml. Accessed 31 October 2012.

21. B. A. Wagner, Physical Properties and Spectra, in *Chemical Information for Chemists: A Primer*; ed. J. N. Currano and D. L. Roth, Royal Society of Chemistry, Cambridge, 2012, p. 146.

II
The Primary Literature

CHAPTER 2

Non-Patent Primary Literature: Journals, Conference Papers, Reports, Abstracts and Preprints

DANA L. ROTH

California Institute of Technology, Millikan 1-32, Pasadena, CA 91125, US
Email: dzrlib@library.caltech.edu

2.1 INTRODUCTION TO THE NON-PATENT PRIMARY LITERATURE

Successful research efforts, in either academe or industry, require an awareness of the state of the art. This is necessary in order to identify previous research, to develop an understanding of the research problem, and to avoid unnecessary duplication of work. In addition, maintaining an on-going awareness of current literature in your area of chemistry, is crucial to career advancement. This is true for graduate students, whose theses must be based on unique research results, as well as for senior researchers who are able to secure grant funding and pursue exclusive technology rights only for original work.

Since there are currently ∼10 000 journals publishing 'chemistry' articles, the historical pattern of perusing the contents pages of a few crucial journals and communicating with colleagues must be supplemented by database searching. There are a variety of databases that can quickly provide a review of recent developments and also offer current awareness alerts.

Chemical Information for Chemists: A Primer
Edited by Judith N. Currano and Dana L. Roth
© The Royal Society of Chemistry 2014
Published by the Royal Society of Chemistry, www.rsc.org

As regards current awareness, it is important to recognize that both SciFinder and PubMed index basic bibliographic data, abstracts and cited references (if available) from articles as soon as they are posted to some journal websites, far in advance of their formal publication in a journal issue (*i.e.*, complete with volume numbering and pagination).

This is in contrast with Web of Science, Scopus and Reaxys, which only index articles after they are formally published, which is also when SciFinder and PubMed provide full indexing.

2.2 JOURNALS

2.2.1 Introduction

While the scientific journal literature[1] has largely completed a conversion from print to electronic format, there is currently a dichotomy faced by researchers accessing the primary journal literature. Those with convenient access to university-subscribed resources are able to display or download full text articles through Google/Google Scholar or *via* indexing and abstracting services that make use of link resolvers, such as SFX.[2]

Researchers, primarily at small companies lacking subscribed resources, however, have several viable options: obtain personal journal subscriptions or copies of articles directly from publishers, or subscribe to pay-per-view services, such as DeepDyve,[3] which offer both rental and purchase options.

Fortunately, for biomedical chemistry researchers, the National Library of Medicine developed PubMed Central[4] in 2000 as a full-text archive of freely available biomedical and life sciences journal literature. In 2008, with the adoption of the NIH Public Access Policy,[5] final peer-reviewed journal manuscripts from NIH funded research must be deposited, upon acceptance for publication, in PubMedCentral. Most of the NIH deposited articles, however, have a short (6–12 month) embargo, because of journal publisher policies. The NIH requirement for full public access is no later than 12 months after publication.

In addition, there are quite a number of journals that are published on an open-access basis. These journals are listed in the Directory of Open Access Journals (DOAJ),[6] which is both searchable and browsable at the journal-title level and searchable (author, title, abstract *etc.*) at the article level.

This access dichotomy has resulted in an on-going struggle between readers and journal publishers over the question of subscription access and open access.

2.2.2 Open Access

The question of open access[8a,b] originally developed from the perceived need for patients to have full access to the medical literature. This may have resulted from a general ignorance that journal article abstracts are freely available from either publisher websites or freely accessible databases, such as PubMed and Google/Google Scholar.

Responsible publishers, whether *via* subscriptions or open access, provide essential added value to author manuscripts (*i.e.*, editorial acceptance, peer review, copy editing, electronic formatting and long-term storage), and providing these services obviously requires adequate compensation.[9a,b] The question then becomes whether dissemination of scientific results through journal articles is best served by subscription access, requiring either a personal or institutional (reader-pays) subscription, or by open access (author pays), which is dependent on author and/or funding agency payment for the publication services.

Historically, many U.S. society publishers offered a hybrid model that combined reasonable subscription prices with author page charges. This model continues today for a few titles (*e.g.*, *Phys. Rev. Letters, J. Biol. Chem.*). However, for many society publishers (*e.g.*, ACS) this model has been virtually undone by the rapid rise of commercially published journals that offer increasingly expensive institutional subscriptions, without assessing author page charges.

One interesting option, first offered by the American Physical Society and now adopted by the American Astronomical Society, is to allow on-site public and high school library access to complete archives of their journals. The APS goal 'is to provide access to everyone who wants and needs our journals'.[9] This access is available at no cost, providing that libraries accept a site license and provide the IP addresses for their public-use computers.

There are essentially two versions of open access. 'Gold' is defined as open access to the version of record, the final published version available directly from the journal publisher. The cost of Gold open access is the author's responsibility, and varies considerably, from $\sim$$1000 for society-published journals (*e.g.*, PNAS, ACS members at subscribing institutions) to $\sim$$3000 for society-published journals (*e.g.*, ACS non-member at non-subscribing institutions) and commercially published journals (*e.g.*, Elsevier, Springer). The risks of Gold open access are: (1) the 'culling of the scientific record', by making it difficult for those who are not research funded to publish their work; and (2) the concern that, by paying to publish, acceptance rates may rise at the expense of quality. This is a concern that is exacerbated by the reported existence of predatory publishers operating essentially as vanity presses.[10]

One significant 'Gold' variation is the American Chemical Society, which gives authors a unique AoR (Author of Record) URL link allowing 50 downloads during the first year after publication and unlimited downloads in succeeding years. These URLs can be posted on an author's webpage or institutional repository.

'Green' open access is based on authors, or their surrogates, posting their final peer-reviewed and accepted-for-publication manuscripts on a personal or institutional website. Author participation in providing this access, while generally permitted by responsible publishers, has been difficult to achieve in the absence of institutional mandates.

While there is continuing pressure for the provision of Gold open access, especially for publications resulting from government-funded research, further discussion is beyond the scope of this chapter. Additional information is available in both The STM Report and the Finch Report.[11]

2.2.3 Indexing and Abstracting Services

Journal articles are indexed and searchable in both subscription databases, *e.g.*, SciFinder and Web of Science (WoS); freely available databases, *e.g.*, PubMed and Google/Google Scholar; and most publisher websites, *e.g.*, SpringerLink and ACS Publications.[12] Each database has unique criteria for both journal selection and subject coverage. ULRICHSWEB is a global serials directory that provides detailed information on over 300 000 periodicals, including abstracting and indexing coverage.[13]

SciFinder is the electronic version of Chemical Abstracts, which ceased publication, in print, at the end of 2009. SciFinder consists of multiple files, with the CAplus file[14] containing bibliographic data, indexing terms, abstracts and related data, for each record. The CAplus file, which currently has > 36 million records, is unique in that it extends the definition of chemistry beyond the traditional areas of: analytical; applied; biological; inorganic; macromolecular; organic; and physical chemistry to include the chemical aspects of astronomy, structural biology, education, engineering, economics, geology, history, mathematics, medicine and physics.

Chemical Abstracts has a long history of indexing a wide variety of publications: journal articles, chemical patents, conference papers, dissertations, technical reports, technical disclosures and chapters from monographs. In 1994, CA began abstracting 1580 core chemical journals[15] cover-to-cover, providing 'in-process' records prior to full indexing. Thus, bibliographic and abstract information is generally available in CAplus within 8 days of receipt, with full indexing completed within 30–90 days. This explains why an author or keyword search will retrieve some very recent papers, while a substance search will not. For patents from the primary patent offices (US, EP, WO, JP, UK, FR and DE) bibliographic and abstract (machine translations for JP and DE) information is available within two days of publication and full indexing is completed within 27 days.[16] CAplus has also extended the contents of the print Chemical Abstracts, to include records for over 224 000 journal articles, abstracts and patents published before 1907.[17]

The CAS Source Index (CASSI)[18] is a cumulative record of publications indexed and abstracted in SciFinder. It is freely searchable by title, abbreviation, ISBN or ISSN. CLICAPS[19] is a university library catalog that has journal records cataloged according to CASSI, with full title, extensive abbreviations, and succeeding and preceding titles.

Web of Science[20] is a journal-article database containing bibliographic data, indexing terms, abstracts, and related data. Web of Science originated the concept of cited reference searching and offers both 'Cited Reference Searching' (as they appear in the various journal article reference lists) as well as 'Cited By' links (to correctly cited references that are linked to their respective Web of Science records). Selection of the journals indexed in Web of Science[21] is generally based on each journal's Impact Factor.[22] The Conference Proceedings Citation Index[23] and the Book Citation Index[24] are also available, within Web of Science, in a fully integrated package.

PubMed[25] is a freely available journal-article database on Medline, a bibliographic database with a subject focus on all aspects of medicine, and its related sciences and engineering. Medline[26] currently indexes ~5600 international journals and has ~20 million records dating from 1946, and is adding ~3000 records each day. Medline records are extensively indexed with NLM Medical Subject Headings (MESH),[27] a controlled vocabulary thesaurus. In addition to Medline, PubMed also contains in-process records prior to Medline indexing, and records for out-of-scope articles in Medline journals (*i.e.*, general science and chemistry) whose life science articles are indexed with MESH for Medline. When searching PubMed, 'Advanced Search' is strongly recommended. After a search, clicking on the title of an article, in the initial display, displays the complete record, which includes the Abstract, Related citations, Cited by (PubMedCentral) articles, *etc.* A SciFinder search also includes results from a 'Medline' database, which is essentially the same as PubMed.[28]

Google/Google Scholar[29] are freely accessible web databases that index a wide range of scholarly literature but have a variety of unstated limitations. As opposed to other databases, Google/ Google Scholar provide no listing of the publications that are indexed. Use of both Google Scholar's Advanced Scholar Search and Scholar Preferences is strongly recommended. The list of results is ranked according to a variety of factors,[30] and provides 'Cited By' links to references in the database, related articles, and links to multiple versions (records of the article in a variety of databases, *e.g.*, PubMed, NASA ADS, JSTOR). Google is very often a good resource for the full text of specific articles, which may be retrieved by searching with full article titles.

In contrast to the comprehensive A&I services, whose indexing is based on bibliographic and abstract information supplied by publishers, many publishers also provide databases that offer full-text searchable content limited to their publications.

Science Direct[31] is a representative example of a major commercial publisher database. It currently provides searchable access to nearly 10 million articles and chapters (adding about 500 000 records per year), from over 2500 journals and 10 000 books. Book and journal titles are both browsable and searchable, and the article/chapter contents are full-text searchable. While the full text requires subscription access, the abstracts, figures and tables are freely displayed. From the abstract display, an article's corresponding record in Scopus (Elsevier's equivalent to Web of Science) is available. The 'View Record in Scopus' displays the two most recent citing articles, and the five most recent articles by the original article's authors. Other commercial publisher databases include SpringerLink and the Wiley Online Library (which includes journals published by the Gesellschaft Deutscher Chemiker, ChemPubSoc Europe and the Asian Chemical Editorial Society).

ACS Publications[32] is representative of non-profit society publishers. It currently provides full-text searchable access to over 1 million articles from journals, and chapters in its ACS Symposium Series and Advances in Chemistry. The ACS Pubs webpage provides an alphabetical listing of titles and a default search box for full-text searching of keywords, citations, DOI, or 'within a specific subject'. 'Advanced Search' offers Boolean, phrase, wildcard, stemming and limiting a search to a specific journal title. Full-text article access requires a subscription but abstract display is free. ACS has recently announced new member benefits, which include online access (limited to 48 hours from initial article request) to any 25 articles from ACS Journals, ACS Symposium Series, and C&EN

Archives. Another important society publisher is the Royal Society of Chemistry, which offers full-text searching for its journal articles and book chapters.[33]

2.2.4 Current Awareness

Awareness of current research results is an essential component of any research program. In the print journal/database era, librarians routinely routed journal issues and/or photocopies of their contents, and researchers routinely visited the library to peruse recent issues and/or maintained personal subscriptions. Since then, with the enormous increase in publication and the transition to electronic journals and databases, and widespread use of laptops and mobile devices, these practices have been replaced with email or RSS alerts from online databases and/or directly from journal publishers.

The following databases provide users with options to save search statements and have them run automatically on a regular schedule. The journal publisher websites offer e-mail announcements for each new journal issue or book volume.

SciFinder
https://scifinder.cas.org

SciFinder offers an 'add KMP Alert' link, which can be activated from any 'references' list. Clicking this link opens a dialog box for entering: Title, Description, Duration, and Frequency (weekly or monthly), as well as an option to exclude previously retrieved references (since, for example, many ACS articles are indexed ahead of print, and require updating).

Web of Science
http://apps.webofknowledge.com

Web of Science (personal login required) offers a variety of search 'alerts', (*e.g.*, author, topic, address, or cited reference). Following a search, a search 'alerts' is initiated by clicking 'Search History' (top menu bar). The last query on the 'Search History' page will be used to generate the 'alert'. Then click 'Save History/Create Alert'. After editing the fields on the 'Save Search History' page, check-off 'Send Me E-mail Alerts' and click 'Save'. A 'saved' search strategy cannot be edited. It must be cancelled and replaced with a new search strategy. Web of Science also offers cited reference 'alerts'. Each WoS record has a 'Create Citation Alert' tab that, when clicked, will result in an e-mail alert being sent out each time this article is cited by a newly published article. Cited reference 'alerts' are very useful for identifying research articles that have recently cited a previously published article.

PubMed
http://www.ncbi.nlm.nih.gov/sites/entrez

PubMed offers 'My NCBI', which can be used to save search strategies, search results and a bibliography, and has an option to automatically run saved searches and email the results. First click on the 'My NCBI' tab and register. After signing in, click the 'My NCBI' tab, then run a search and click 'Save search'. The web page screens that follow walk the user through the process that results in monthly, weekly or daily updates.

Google Alerts
http://www.google.com/alerts

'Google Alerts' are based on the most recent Google results (web, news, blogs, videos, discussion or books). Email updates are based on entering a 'Search query', which can be set to search 'Everything' or limited to a specific category. Alerts are sent: 'As-it-happens'; Once a day; or Once a week and can be limited to either 'Only the best results' or 'Everything'.

Science Direct
http://www.sciencedirect.com/

Science Direct offers 'My alerts' (following free registration). 'My alerts' choices include: Search alerts (define and run a search, then save the search as a Search Alert); topic alerts (select a pre-defined topic, *e.g.*, electrochemistry, organic chemistry *etc.*); and volume/issue alerts (for individual journal issues or book volumes, as published).

ACS Publications
http://pubs.acs.org/

The ACS Publications home page offers 'e-alerts' that are either ASAP Alerts (notification when individual articles are released on the web), or TOC Alerts (notification when the table of contents for specific journal issues are posted). There is a 'Log In' button that leads to the ACS ID registration page (ACS membership is NOT required). After logging in, select the journals of interest, the desired alert frequency, and preferred email format.

2.2.5 Journal Impact

Given the enormous number of research journals currently being published, it is important to have some measure of their quality. In this regard, the journal impact factor (IF) is frequently used by researchers as a criterion for selecting appropriate journals for research-publication submission. The IF was first introduced by Eugene Garfield,[34] at the Institute for Scientific Information, in the Science Citation Index and is currently calculated on an annual basis for the journals indexed in the Web of Science, appearing in June of the following year in the Journal Citation Reports (subscription required).[35]

A journal's ISI Impact Factor[36] is the average number of cites, in a given year, to the articles published in that journal during the two preceding years. For example, the *Journal of the American Chemical Society* averaged 9.907 cites in 2011 to each article published in its 2009 and 2010 issues. For example, to derive the Impact Factor for the *Journal of the American Chemical Society*:

$$\frac{64,111 \text{ cites to } (2009 + 2010) \text{ JACS articles}}{6,471 \ (2009 + 2010) \text{ JACS articles}} = 9.907 \ (2011 \text{ JACS IF})$$

ISI also provides a five-year Journal Impact Factor, which is the average number of cites, for example, in 2010 to articles published in the 2005–2009 issues.

While generally valid, this averaging process can lead to some interesting anomalies. For example, *Acta Crystallographica. Section A*, which had an Impact Factor of ~ 2 from 2006 through 2008, saw a sudden increase to ~ 50 in 2009 and 2010, as a result of one very heavily cited article published in 2008.[37] Another example, of a rapid increase in a journal's Impact Factor, resulted from a single review article that cited over 400 recent articles in *Cell Transplantation*, almost doubling its Impact Factor (from 3.5 to 6.2).[38]

The Arizona State University Libraries has posted a 'Citation Counts' library guide for Web of Science, Google Scholar, and other alternatives.[39] 'Citation Counts' are defined simply as the number of times a publication is cited by others. This measure is complicated by the fact that none of citation sources index all relevant publications. This very comprehensive guide provides both a listing of the many sources available for creating citation counts and a detailed explanation of their coverage and method of searching.

Recent competitors to the ISI Impact Factor are the SNIP (Source-Normalized Impact per Paper) and the SCImago Journal Rank (SJR). These databases attempt to normalize citing references across subject fields, thus allowing comparison of journals irrespective of their primary subject content.[40]

SNIP[41] uses the ratio of citing references to the journal's citation potential (which is based on the citation pattern of the subject field), thus allowing comparison of diverse journals; whereas SJR[42] is based on the idea that the quality and reputation of the citing journal should affect the value of its citations.

SNIP and SJR values are publicly accessible through JournalM3-trics[43] and are also integrated into the Scopus Journal Analyzer for its subscribers.[44] They provide data on ~ 18 000 journals, as well as proceedings and book series.

Another approach is reflected in the Eigenfactor Score, which is derived from the total number of citations to articles published in the previous five years in a given Journal Citation Report. The raw numbers are then refined so that journal self-citations are eliminated and citations from highly cited journals are given more significance.

Thomson Reuters includes Eigenfactor Scores in the Web of Knowledge Journal Citation Reports along with the corresponding Article Influence Score. From the WoK JCR website:

'The Article Influence determines the average influence of a journal's articles over the first five years after publication. It is calculated by dividing a journal's Eigenfactor Score by the number of articles in the journal, normalized as a fraction of all articles in all publications. This measure is roughly analogous to the 5-Year Journal Impact Factor in that it is a ratio of a journal's citation influence to the size of the journal's article contribution over a period of five years.... The mean Article Influence Score is 1.00. A score greater than 1.00 indicates that each article in the journal has an above-average influence. A score less than 1.00 indicates that each article in the journal has below-average influence.'[45]

Eigenfactor (EF) and Article Influence (AI) scores are also available at the eigenFACTOR.org website.[46] Care must be taken when searching with journal titles, since it is not obvious to which year the data refers. Clicking on the EF or AI values, however, currently displays graphical data plotted for 1997–2010 for both EF and AI, along with graphical data on cost-effectiveness for 2007–2010.

For comparison purposes, a selection of titles, and their 2011 (JCR) values are given in Table 2.1.

When comparing journals, it is very important to recognize that review journals are much more heavily cited and should not be

Table 2.1 2011 impact metrics for four widely read, general chemistry journals.

Journal	ISI IF	SNIP	SJR	EF	AI
Angew Chem IE	13.455	3.963	1.123	0.51393	3.370
J Am Chem Soc	9.907	4.058	1.117	0.81677	2.792
Chem Commun	6.169	1.942	0.507	0.24077	1.547
Chem-Eur J	5.925	1.768	0.455	0.16880	1.527

Table 2.2 2011 impact metrics for five review journals, covering various areas of chemistry.

Journal	ISI IF	SNIP	SJR	EF	AI
Chem Rev	40.197	15.865	3.791	0.21464	13.305
Ann Rev Biochem	34.317	10.131	7.805	0.05695	19.743
Chem Soc Rev	28.760	8.666	2.493	0.13670	8.069
Ann Rev Phys Chem	14.130	6.295	1.614	0.01687	7.467
Coord Chem Rev	12.110	4.195	0.789	0.04170	3.206

compared to research journals. Although *Angew Chem IE* is primarily a research journal, its publication of review articles is one factor of its significant ISI IF.

Examples of important review journals, and their 2011 (JCR) data are given in Table 2.2.

2.2.6 Article Impact

The most commonly used metric for article impact is the h-index,[47] named for its developer J.E. Hirsch). The h-index, shown in a Web of Science Citation Report by an orange colored line, is based on a listing of publications ranked in descending order by the Times Cited count. To discount the effect of less significant publications, the articles in this ranking should be edited (*e.g.*, to eliminate translations) and then limited to research and review articles. The numerical value of 'h' then is the number of articles (h) with at least (h) citations.

The h-index, however, tends to discount both highly and poorly cited articles. It rewards an author's longevity and productivity but penalizes authors with a relatively small number of highly cited articles (*e.g.*, A. Einstein [h = 53] and R.P. Feynman [h = 33]), whose h-index is small compared with much more prolific authors (*e.g.*, G. Whitesides [h = 160]) but whose average citations per article are 154, 536, and 120, respectively. Please note that the h-index value, calculated from Web of Science data, is also a function of both the citation databases and the time-span, which can also be refined prior to the search to ensure an accurate and comparable set of publications, when comparing multiple authors.

It can even be used to compare articles on a particular subject published in different journals, to help select high impact publishing venues.

The concept of altmetrics,[48] for individual articles, provides a more immediate measure that supplements counting citations, by aggregating data from multiple online sources (*e.g.*, tweets, blogs, bookmarks, citation counts *etc.*).

For example, Total Impact[49] currently tracks a wide range of research output, *e.g.*, papers, datasets, software, preprints and slides. Searchable terms include; DOI, PMID and URL and search results include HTML views, PDF views, citations and cited-in. On the Total Impact site, click on 'add to collection' and then click 'get my metrics' to see a sample search.

Harzing's Publish or Perish[50] is a downloadable software program that retrieves and analyzes academic citations, using Google Scholar to obtain statistics that include: citations to and from journal articles, book chapters, patents, conference papers and technical reports (*i.e.*, anything in Google Scholar).

Google Scholar Citations(GSC)[51] provides a personalized citation analysis (following registration). It is possible to search for authors who have made their profile public in either GSC or Google Scholar. Google Scholar Citations is updated continually, allowing tracking of citations to your publications, graphing your citations over time and computing citation metrics.[52] This data may be compromised, however, by 'Ghost Authors'[53] that are variant forms of the same paper with citations that are mismatched by Google's crawler and parser programs.

A Google Scholar h-index Calculator is provided as a Firefox Add-On[54] but the results are problematic when refining search results to a specific author.

Several websites that host journal articles provide download data for individual articles that often far exceeds the citation counts in Web of Science or Scopus. *Sensors & Transducers Journal*, which posts a rolling list of the 25 Top List of Most Downloaded Articles[55] for the previous month, with a 'Previous' link at the bottom of the page for additional data. The Jefferson Digital Commons[56] sends out monthly reports to authors, providing, for example, the total number of electronic downloads of papers published in the Special Libraries Association's Sci-Tech News. This service is based on the 'bepress Download Totals'.[57]

Since March 2009, the Public Library of Science (PLoS)[58] has provided 'Article-Level Metrics (ALMs)', which currently include:

- article usage statistics (HTML page views, and PDF and XML downloads);
- citations (CrossRef, PubMed Central, Scopus, Web of Science, Google);

- social networks (Facebook and Mendeley);
- blogs and media coverage (Google); and
- PLoS Readers (Rate this article, Comments & Notes).

Many of these alternative methods for citation counts and journal rankings are compiled on a library guide[39] that links to a variety of options and includes 'terms and definitions', 'tutorials' and 'further reading'.

2.3 CONFERENCE PAPERS, REPORTS, ABSTRACTS AND PREPRINTS

2.3.1 Introduction

Conference papers, reports, abstracts and preprints are commonly described as the 'Gray Literature', a term used by librarians for literature that was, for many years, often difficult to locate or obtain using commercial indexing and abstracting services. This situation has generally improved with the availability of the Conference Proceedings Citation Index (either as a separate subscription or in combination with Web of Science), and the resources provided by the British Library,[59] which rightly claims to have the most comprehensive collection of publically available conference proceedings and reports.

A Caltech online library guide listing both open access conference paper websites (*e.g.*, Elsevier Procedia – Chemistry, IOP Conference Series: Materials Science and Engineering) and indexing/abstracting databases (*e.g.*, U.S. government, British Library) are available.[60]

The E-Print Network[61] is a full-text searchable gateway to a worldwide collection of gray literature websites and databases of interest to DOE researchers. It contains preprints, reprints, technical reports and conference publications; primarily in physics, but also including chemistry, life sciences, materials science and energy research. It provides access to over 35 000 websites and databases around the world, with over 5.5 million entries.

A comprehensive review article with links to free online science, technology and engineering resources is available.[62]

The National Technical Information Service (NTIS), the National Aeronautics and Space Administration (NASA) and the U.S. Department of Energy (DOE) all provide freely available online indexing/abstracting services for research reports funded by the U.S. government.

Preprints, however, are not an established culture in chemistry. The American Chemical Society's journal editorial policy, in contrast with the ACS Symposium Series, treats preprints as prior publication. If a submitted article is found to have been posted on a preprint server, it will be withdrawn from consideration for publication in any of the ACS journals.

The ACS Symposium Series volumes, however, are often based on revised preprints that originally appeared in the following publications:

> *Preprints ... Division of Environmental Chemistry*
> *Preprints ... Division of Fuel Chemistry*
> *Preprints ... Division of Petroleum Chemistry*
> *PMSE Preprints ... Division of Polymeric Materials Science & Engineering*
> *Polymer Preprints ... Division of Polymer Chemistry*

Elsevier's Science Direct provided a preprint server but due to a lack of contributed articles stopped accepting new submissions on May 24, 2004.

2.3.2 Indexing and Abstracting Services

The following entries provide descriptions of resources that include chemistry-related information. While SciFinder indexes conference papers, reports, some arXiv and CERN preprints and ACS Meeting Abstracts, chemists interested in comprehensively searching for government sponsored research should take advantage of the specialty databases described below. Since each database will provide unique results, it is probably important to experiment with each one for maximum results.

ACS. Technical Programming Archive of Past National Meetings
http://portal.acs.org/portal/PublicWebSite/meetings/nationalmeetings/programarchive/

This ACS website provides both browsable and searchable publically available access for each individual National Meeting since the 227th in March, 2004.

arXiv
http://arxiv.org

The arXiv electronic preprint archive[63] covers the fields of mathematics, physics, astronomy, computer science, quantitative biology, statistics and quantitative finance. While arXiv preprints are only of tangential interest to chemists, SciFinder has indexed all preprints (nearly 280 000), beginning in 2000, in the following subject areas: Astrophysics, Condensed Matter, General Relativity and Quantum Cosmology, High Energy Physics, Nuclear Experiment, Nuclear Theory, Physics, Quantitative Biology and Quantum Physics.

BASE: Beilefeld Academic Search Engine
http://www.base-search.net/

BASE provides access to document servers that comply with the specific requirements of academic quality and relevance. It currently contains nearly 35 million documents (both journal articles and technical reports) from over 2000 content sources.

Chemical Abstracts (SciFinder & CAplus on STN)
http://www.cas.org/expertise/cascontent/caplus/confcov.html

The CAplus file, in STN and SciFinder, has a long history of indexing conference papers from journal issues, separately published conference monographs, ACS National Meeting Divisional Symposia, as well as NTIS reports and, since 1995, have indexed and provided the full text of ACS National Meeting Abstracts, while maintaining its policy of not indexing publications simply consisting of abstracts. The 'Explore References' option in SciFinder allows selection of all or individual document types: Conference (which includes ACS Abstracts), Preprint, Report.

Energy Citations Database (ECD)
http://www.osti.gov/energycitations/

The ECD provides free public access to over 2.5 million report citations from 1943 to the present. It includes a full range of sci/tech research results of interest to the DOE covering the report literature, conference papers, journal articles, books, dissertations and patents. It includes the Nuclear Science Abstracts (1948-1976), and the ERDA/Energy Research Abstracts (1976–1994) databases.

Google
http://www.google.com

Confex.com provides for collection and management of presentations from hundreds of conferences annually. These presentations are indexed by Google and searchable by adding (confex.com and a sponsor acronym), to keywords and/or author names. For example, a search for

Harry Gray's paper on Metal ligand triple bonds and the oxo wall, which was presented at the 2009 ACS meeting could be searched as:

Harry Gray triple 2009 (confex.com and ACS) which retrieves: http://oasys2.confex.com/acs/237nm/techprogram/P1231639.HTM

Information Bridge: DOE Scientific and Technical Information
http://www.osti.gov/bridge/
Information Bridge provides freely available access to nearly 300,000 full text documents and citations for research reports funded by the DOE, from 1991 +. Subjects include 'chemistry', 'materials', 'renewable energy' *etc.*

International Nuclear Information System (INIS)
http://www.iaea.org/inis/

INIS is the successor to the Atomindex, which first appeared on 1970. INIS is focused on peaceful uses of nuclear science and technology. The keyword default search is supplemented with an Advanced Search which allows both 'include' and 'exclude' options.

NASA Technical Reports Server (NTRS)
http://ntrs.nasa.gov/search.jsp

The NTRS database has indexing/abstracting information for over 500 000 aerospace citations (including 'chemistry and materials') with over 200 000 that are online full-text. In addition, there is information on over 500 000 images and videos. NTRS integrates the NACA (National Advisory Committee for Aeronautics) citations and reports (1916–1958) and the NASA (National Aeronautics and Space Agency) citations and documents (1958–present) with the NIX (NASA Image eXchange) collection of citations which has links to images, photos, movies and videos.

NTIS.gov
http://www.ntis.gov/

In addition to the NTIS bibliographic database which contains over 2 million records, this website also offers RSS current awareness feeds for nearly 40 subject areas, including 'Chemistry', 'Combustion', 'Energy' and 'Materials Sciences'. In addition, NTIS is the source for more than 150 subscriptions produced by U.S. government agencies.

Science Conference Proceedings
http://www.osti.gov/scienceconferences/

This is a distributed portal that provides access to sci/tech conference proceedings and papers published by both professional societies and national laboratories) that are of interest to the Department of Energy. Search options include: Select All or limited to specific societies (*e.g.*, ACS, AOCS), national laboratories (*e.g.*, EPRI, NIST) or the Energy Citations Database.

Science Research.com
http://www.scienceresearch.com/scienceresearch/advancedsearch.html

ScienceResearch.com is a free, publicly available federated search engine. Federated searches are defined by their ability to submit search terms – in real time – to a pre-defined collection of databases. Results are then collated, ranked, and de-duplicated. Science Resarch.com offers a variety of text search and subject options.

2.3.3 Current Awareness

Most databases provide alert services which provide notification of newly added content. In addition, while conference papers often appear in regularly published journal issues, others are appearing in online (often open-access) monographic volumes. Many of these separately published volumes appear in publisher series, which provide new volume alerts, such as:

AIP Conference Proceedings (Subscription required)
http://proceedings.aip.org/browse/new_titles

AIP (American Institute of Physics) Conference Proceedings covers worldwide scientific meetings which provide topical status reports on a wide variety of subjects.

IOP (Institute of Physics) Conference Series: Earth and Environmental Science
https://conferenceseries.iop.org/theme/ees

An Open Access publication that complements a community web site, environmentalresearchweb.org and an open access journal, Environmental Research Letters.

IOP (Institute of Physics) Conference Series: Materials Science and Engineering
https://conferenceseries.iop.org/theme/mse

An Open Access publication with conference papers on material science, physics, chemistry and engineering.

Journal of Physics: Conference Series
https://conferenceseries.iop.org/theme/jpcs

An Open Access publication with conference papers on physics, physical chemistry and biophysics.

Elsevier's Procedia Series
http://www.sciencedirect.com

A series of open-access conference proceedings on a wide variety of subjects, including: Energy Procedia, Procedia Chemistry, Procedia Environmental Sciences and Procedia Food Science.

EJP (European Journal of Physics) Web of Conferences
http://www.epj-conferences.org/

Open-access proceedings in pure and applied physics, including materials science, physical biology, physical chemistry, and complex systems.

ACS (American Chemical Society Symposium Series [subscription])
http://pubs.acs.org/series/symposium

The ACS Symposium Series contains peer-reviewed papers based on presentations at ACS National Meeting symposia. The series covers a broad range of topics including: food chemistry, chemical education, organic chemistry, polymer chemistry, materials science *etc*. The first chapter in every volume of the series is freely available to view. Availability of all other chapters is based on institutional subscription or membership of the ACS.

REFERENCES

1. Scientific journal. *Wikipedia, the Free Encyclopedia*, http://en.wikipedia.org/wiki/Scientific_journal Accessed April 2, 2013.
2. SFX – the OpenURL link resolver and much more, http://www.exlibrisgroup.com/category/SFXOverview, Accessed April 2, 2013.
3. DeepDyve, http://www.deepdyve.com/, Accessed April 2, 2013.
4. PubMed Central. *Wikipedia, the Free Encyclopedia*, http://en.wikipedia.org/wiki/PubMed_Central, Accessed April 2, 2013.
5. NIH Public Access Policy, http://publicaccess.nih.gov/, Accessed April 2, 2013.
6. DOAJ – Directory of Open Access Journals, www.doaj.org/, Accessed April 2, 2013.
7. J. M. Drazen, G. D. Sprouse, J. W. Serene. Should Research Be More Freely Available? *New York Times*, January 23, 2012, http://nyti.ms/edletter0112, Accessed April 2, 2013.

8. M. Leptin. Open Access—Pass the buck, *Science*, 2012, **335**, 1279. 10.1126/science.1220395; Open access *Wikipedia, the Free Encyclopedia*, http://en.wikipedia.org/wiki/Open_access, Accessed April 2, 2013; Scholarship 2.0: An idea whose time has come, http://scholarship20.blogspot.com/2012/04/open-access-will-open-new-ways-to.html, Accessed April 2, 2013; Open access journal. *Wikipedia, the Free Encyclopedia*, http://en.wikipedia.org/wiki/Open_access_journal, Accessed April 2, 2013.

9. (a) APS Online Journals Available Free in U.S. Public Libraries, http://librarians.aps.org/public-access-announcement, Accessed April 2, 2013; (b) APS Online Journals Available Free in U.S. High Schools (February 9, 2011), http://pre.aps.org/highschool-access-announcement, Accessed April 2, 2013.

10. Beall's List of Predatory, Open-Access Publishers, http://metadata.posterous.com/83235355, Accessed April 2, 2013.

11. The STM Report: An overview of scientific and scholarly journal publishing, http://www.stm-assoc.org/2009_10_13_MWC_STM_Report.pdf, Accessed April 2, 2013; Finch Report: Accessibility, sustainability, excellence: how to expand access to research publications. http://www.researchinfonet.org/wp-content/uploads/2012/06/Finch-Group-report-FINAL-VERSION.pdf, Accessed April 2, 2013.

12. PubMed, http://www.ncbi.nlm.nih.gov/pubmed/, Accessed April 2, 2013; Google, http://www.google.com/, Accessed April 2, 2013; Google Scholar, http://scholar.google.com/, Accessed April 2, 2013; About Google Scholar, http://scholar.google.com/intl/en/scholar/about.html, Accessed April 2, 2013; SpringerLink, http://www.springerlink.com/, Accessed April 2, 2013; ACS Publications, http://pubs.acs.org/, Accessed April 2, 2013.

13. ULRICHSWEB: Global Serials Directory, http://ulrichsweb.serialssolutions.com/, Accessed April 2, 2013.

14. SciFinder CAplus file, http://www.cas.org/expertise/cascontent/caplus/index.html, Accessed April 2, 2013.

15. CAplus Core Journal Coverage List, http://www.cas.org/content/references/corejournals, Accessed April 2, 2013.

16. CAS Coverage of Patents, http://www.cas.org/content/references/patentcoverage, Accessed April 2, 2013.

17. CAplus – Pre-1907 Coverage, http://www.cas.org/content/references/capluspre1907, Accessed April 2, 2013.

18. CAS Source Index (CASSI) Search Tool, http://cassi.cas.org/search.jsp, Accessed April 2, 2013.

19. CLICAPS (ETH Zurich) OPAC, http://www.clicaps.ethz.ch/en/, Accessed April 2, 2013.

20. Web of Science, http://wokinfo.com/products_tools/multidisciplinary/webofscience/, Accessed April 2, 2013.
21. Thomson Reuters Master Journal List, http://ip-science.thomsonreuters.com/mjl/, Accessed April 2, 2013.
22. Impact factor. *Wikipedia, the Free Encyclopedia*, http://en.wikipedia.org/wiki/Impact_factor, Accessed April 2, 2013.
23. Conference Proceedings Citation Index (CPCI), http://thomsonreuters.com/products_services/science/science_products/a-z/conf_proceedings_citation_index/, Accessed April 2, 2013.
24. Book Citation Index (BCI), http://wokinfo.com/products_tools/multidisciplinary/bookcitationindex/, Accessed April 2, 2013.
25. PubMed Fact Sheet, http://www.nlm.nih.gov/pubs/factsheets/pubmed.html, Accessed April 2, 2013.
26. Medline Fact Sheet, http://www.nlm.nih.gov/pubs/factsheets/medline.html, Accessed April 2, 2013.
27. Medical Subject Headings (MESH) Fact Sheet, http://www.nlm.nih.gov/pubs/factsheets/mesh.html, Accessed April 2, 2013.
28. What's the Difference Between MEDLINE® and PubMed, http://www.nlm.nih.gov/pubs/factsheets/dif_med_pub.html, Accessed April 2, 2013.
29. Google Scholar. *Wikipedia, the Free Encyclopedia*, http://en.wikipedia.org/wiki/Google_Scholar, Accessed April 2, 2013.
30. About Google Scholar, http://scholar.google.com/intl/en/scholar/about.html, Accessed April 2, 2013.
31. Science Direct (Elsevier), http://www.sciencedirect.com/, Accessed April 2, 2013; About Science Direct, http://www.info.sciverse.com/sciencedirect/about, Accessed April 2, 2013; Science Direct (Intellogist), http://www.intellogist.com/wiki/ScienceDirect, Accessed April 2, 2013.
32. ACS Publications. *Wikipedia, the Free Encyclopedia*, http://en.wikipedia.org/wiki/American_Chemical_Society#Journals_and_magazines, Accessed April 2, 2013.
33. RSC Publishing: Advanced Search, http://pubs.rsc.org/en/search/advancedsearch, Accessed April 2, 2013.
34. E. Garfield. The History and Meaning of the Journal Impact Factor. *JAMA*, 2006, 295, 90. http://jama.jamanetwork.com/article.aspx?articleid = 202114; E. Garfield. The Agony and the Ecstasy—The History and Meaning of the Journal Impact Factor. Presented at the International Congress on Peer Review And Biomedical Publication, Chicago, IL, September 16, 2005. http://garfield.library.upenn.edu/papers/jifchicago2005.pdf.

35. Journal Citation Reports, http://webofknowledge.com/JCR, Accessed April 2, 2013.
36. Impact Factor. *Wikipedia, the Free Encyclopedia*, http://en.wikipedia.org/wiki/Impact_factor, Accessed April 2, 2013.
37. At the time of this writing, G. M. Sheldrick. A short history of SHELX. *Acta Crystallogr.* 2011, 64, 112. (http://scripts.iucr.org/cgi-bin/paper?sc5010) had been cited 21,913 times in the Web of Science.
38. The Emergence of a Citation Cartel. *MetaFilter: community weblog*, http://www.metafilter.com/115956/The-Emergence-of-a-Citation-Cartel, Accessed April 2, 2013.
39. Citation Research (Arizona State University Library), http://libguides.asu.edu/content.php?pid = 11186&sid = 74734, Accessed April 2, 2013.
40. M. E. E. Falagas. Comparison of SCImago journal rank indicator with journal impact factor. *The FASEB Journal*, 2008, **22**, 2623; P. Jacso. Comparison of journal impact rankings in the SCImago Journal & Country Rank and the Journal Citation Reports databases. *Online Information Review*, 2010, **34**, 642, http://www.emeraldinsight.com/journals.htm?articleid = 1876484&show = pdf.
41. About SNIP, http://www.journalmetrics.com/snip.php, Accessed April 2, 2013.
42. About SJR, http://www.journalmetrics.com/sjr.php, Accessed April 2, 2013.
43. JournalM3trics, http://www.journalmetrics.com/, Accessed April 2, 2013.
44. Scopus Journal Analyzer, http://www.info.sciverse.com/scopus/scopus-in-detail/tools/journalanalyzer/, Accessed April 2, 2013.
45. Journal Citation Reports Help, http://admin-apps.webofknowledge.com/JCR/help/h_eigenfact.htm, Accessed April 2, 2013.
46. eigenFACTOR.org, http://www.eigenfactor.org/, Accessed April 2, 2013.
47. h-index. *Wikipedia, the Free Encyclopedia*, http://en.wikipedia.org/wiki/H-index, Accessed April 2, 2013.
48. J. Howard. Scholars Seek Better Ways to Track Impact Online. *The Chronicle of Higher Education*, http://chronicle.com/article/As-Scholarship-Goes-Digital/130482/, Accessed April 2, 2013.
49. Total Impact, http://total-impact.org/, Accessed April 2, 2013.
50. Harzing, A. W. (2007) Publish or Perish, http://www.harzing.com/pop.htm, Accessed April 2, 2013.
51. Google Scholar Citations, http://scholar.google.com/citations, Accessed April 2, 2013; I. F. Aguillo. Is Google Scholar Useful for

Bibliometrics? A Webometric Analysis. *Scientometrics*, 2012, **91**, 343, http://scholarship20.blogspot.com/2012/04/is-google-scholar-useful-for.html.

52. P. Jacso and Google Scholar Author, Citation Tracker: is it too little, too late?, *Online Information Review*, 2012, **36**, 126.
53. P. Jacso, Google Scholar's Ghost Authors, *Library Journal*, 2009, **134**, 26, http://www.jacso.info/PDFs/jacso-google-scholars-ghost-authors.pdf.
54. Google Scholar h-index Calculator, https://addons.mozilla.org/en-US/firefox/addon/scholar-h-index-calculator/, Accessed April 2, 2013.
55. 25 Top List of Most Downloaded Articles, http://www.sensorsportal.com/HTML/DIGEST/Top_articles.htm, Accessed April 2, 2013.
56. Jefferson Digital Commons, http://jeffline.tju.edu/Publishing/jdc.html, Accessed April 2, 2013.
57. The Berkeley Electronic Press Download Totals, http://www.bepress.com/download_counts.html, Accessed April 2, 2013.
58. Public Library of Science (PLoS), http:// www.plos.org/, Accessed April 2, 2013.
59. British Library. The Conference Collections, http://www.bl.uk/reshelp/atyourdesk/docsupply/collection/confs/index.html, Accessed April 2, 2013; British Library, Reports. http://www.bl.uk/reshelp/atyourdesk/docsupply/collection/reports/index.html, Accessed April 2, 2013.
60. Conference Papers/Reports/Abstracts – Publications & Databases, http://libguides.caltech.edu/aecontent.php?pid = 81988, Accessed April 2, 2013.
61. E-Print Network, http://www.osti.gov/eprints/, Accessed April 2, 2013.
62. N. Tchangalova and Francy Stilwell, Search engines and beyond: A toolkit for finding free online resources for science, technology and engineering, *Issues in Science and Technology Librarianship*, Spring 2012, http://www.istl.org/12-spring/internet1.html.
63. arXiv, http://en.wikipedia.org/wiki/ArXiv, Accessed April 2, 2013.

CHAPTER 3
Chemical Patents

MICHAEL J. WHITE

Librarian for Research Services, Engineering & Science Library,
Queen's University, Kingston, Ontario, Canada, K7L 5C4
Email: michael.white@queensu.ca

3.1 HISTORY AND OVERVIEW OF CHEMICAL PATENTS

Chemists have long recognized the value of patents as a source of chemical information. In a speech on the importance of patents to the chemical industry delivered before the American Institute of Chemical Engineers (AIChE) in December 1912, Dr Leo Baekeland, inventor of Bakelite, one of the first commercially successful synthetic plastics, stated "Whoever desires to get posted on the modern literature pertaining to any industrial chemical processes will find that available text-books are many years behind in information as far as novelty and accuracy are concerned; for this reason alone, it is absolutely indispensable to get acquainted with all recent patent literature."[1]

Dr Baekeland's statement is as true today as it was a century ago. In fact, patents have been an important source of chemical information for more than two hundred years. Some of the earliest granted patents were for chemistry-related inventions. The first patent granted in the United States was issued to Samuel Hopkins of Philadelphia on July 31, 1790 for "an improvement, not known or used before such discovery, in the making of Pot Ash and Pearl Ash by a new apparatus and process". In the 18th and 19th centuries in North America, Potash (potassium

Chemical Information for Chemists: A Primer
Edited by Judith N. Currano and Dana L. Roth
© The Royal Society of Chemistry 2014
Published by the Royal Society of Chemistry, www.rsc.org

carbonate) was an important ingredient in the manufacture of fertilizer, soap and black powder. Hopkins, a merchant with far-flung business interests, and his business partners in Montreal also obtained a patent on his process in Lower Canada (Quebec) in 1791.[2] Many consider this to be the first patent granted in Canada.

Many notable chemical inventions over the past two centuries have been patented. In 1844, American inventor Charles Goodyear patented a process for making vulcanized rubber. Plagued by poor business decisions and patent infringement cases, Goodyear died penniless in 1860, but his invention transformed rubber into a material with unlimited practical applications, including protective clothing and tarps, water-proof containers, boots, bicycle and, later, automobile tires.[3] In 1856, British chemist William Perkin patented a process for making artificial dye from coal tar. Perkin called his compound, the first synthetic dye produced on a commercial scale, mauveine, later shortened to mauve.[4] For millennia, dyes had been made from natural substances such as plants, minerals, mollusks and insects, many of which were expensive and difficult to produce. Within a few years, Perkins and other chemists had invented dozens of new artificial dyes, sparking a fashion revolution that swept Europe and North America. Perkin's invention also spurred the development of organic chemistry and interest in industrial applications of chemistry research. One of the first commercially successful pharmaceuticals was acetylsalicylic acid (aspirin), synthesized in 1898 by German chemist Felix Hoffmann and patented around the world by Bayer.

At the time of Dr Baekeland's speech to AIChE, chemical patents were on the verge of taking off in the United States (see Table 3.1). In the early years of the 20th century, the U.S. chemical industry lagged far behind its European counterparts, especially Germany. From 1910 through 1950, only six American chemists received the Nobel Prize in Chemistry, the first being Theodore Richards of Harvard University in 1914. From 1790 through 1900, the U.S. Patent Office issued only a few thousand chemical patents, and from 1900 through 1910, organic chemistry patents accounted for only half a percent of all U.S. patents issued. Over the course of the next several decades, especially after World War I, the number of chemistry patents increased dramatically.[5] For example, by the 1930s, patents on organic compounds accounted for 3.3% of all U.S. patents. The heyday of chemical patents was the 1970s when chemistry patents in the main US patent classes related to chemistry accounted for nearly 15% of U.S. patents issued. The total would be higher if all patents in chemistry-related US classes (260, 423, 424, 514, 518, 520–528, 532–570, and 585) were included. From 1980

Table 3.1 Notable Chemistry Patents of the 19th and 20th Centuries.

Patent	Year	Inventor	Title/Description
US 3,633	1844	Charles Goodyear (US)	Vulcanized rubber
GB 1,864	1856	William Perkin (UK)	Aniline dye (first synthetic dye)
US 88,633	1869	John W. Hyatt (US)	Celluloid (first synthetic plastic)
US RE 11,232	1892	Herbert Henry Dow (US-Canada)	Bromine extraction
US 644,077	1900	Felix Hoffmann (Germany)	Acetylsalicylic acid (Aspirin)
US 942,699	1909	Leo Baekeland (US)	Bakelite (first commercially successful plastic)
US 971,501	1910	Fritz Haber (Germany)	Production of ammonia
US 1,049,667	1913	William Burton (US)	Catalytic cracking of petroleum products
US 1,266,766	1918	Jacques E. Brandenburger (Switzerland)	Cellophane
CA 234,336	1923	Frederick Banting and Charles Best (Canada)	Insulin preparation
US 1,811,959	1931	Julius A. Nieuwland (US)	Neoprene (first synthetic rubber)
US 1,929,435	1933	Waldo L. Sermon (US)	Polyvinyl chloride (PVC)
US 2,130,947	1938	Wallace H. Carothers	Synthetic rubber
US 2,171,765	1939	Otto Röhm (Germany)	Process for the polymerization of methyl methacrylate (Plexiglass)
US 2,752,339	1956	Percy L. Julian (US)	Preparation of cortisone
US 2,825,721	1958	Robert Banks and J. Paul Hogan (US)	High-density polyethylene (HDPE)
US 3,819,587	1974	Stephanie Louise Kwolek	Optically Anisotropic Aromatic Polyamide Dopes and Oriented Fibers (Kevlar®)
US 4,683,202	1987	Kary Mullis (US)	Polymerase Chain Reacton

forward, chemistry-related patents in the main chemistry classes continued to proliferate, with nearly 100 000 granted (Figure 3.1). Although in recent years the percentage of chemistry patents has declined in relation to patents on other technologies, the total number is still formidable. From 2001 through 2012 the USPTO issued more than 160 000 chemistry-related patents.

Recent global chemistry patent trends are similar to the U.S. experience. In Canada from 2008 through 2011, new chemistry-related patent applications accounted for approximately 32% of patent filings.[6] According to the WIPO, in 2010 inventors filed 1.98 million new patent applications worldwide. The top three countries, which account for 62% of filings, are the U.S., China and Japan. Applications for chemistry-related inventions totaled approximately 340 000, or 22.3% of worldwide patent filings. The top three categories were Pharmaceuticals (3.7%), Organic Fine Chemistry (2.8%), and Basic Materials Chemistry (2.5%).[7]

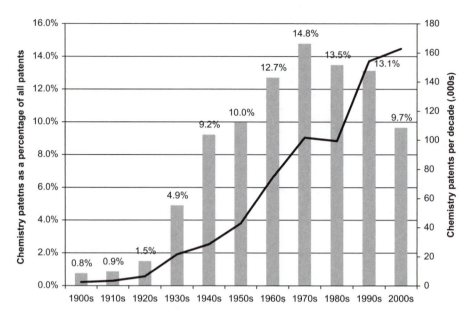

Figure 3.1   Growth of U.S. chemistry patents from 1900 through 2010. This table
consists of patents from USPC classes 260, 423, 424, 514, 518, 520–528,
532–570, and 585.

3.2   BENEFITS OF USING PATENT INFORMATION

Patents are an important source of chemical information for several
reasons. First, as Leo Baekeland recognized a century ago, the information in patents is often more current than the journal and book literature. Even today, despite the proliferation of journals devoted to the
rapid publication of chemical discoveries and the increasing availability
of advanced articles and preprints, patents remain a vital source of
current chemical information. The Chemical Abstracts Service (CAS)
states that 77% of new chemical compounds added to the CAS Registry
are disclosed first in patent applications.[8] Secondly, much of the information contained in patents is not available in other types of publications. Studies have shown that the amount of unique information
contained in patents varies from 10 to 80% depending on the subject
matter.[9] Patents are public documents. Anyone with access to the
internet can search, retrieve and download millions of patent documents
for free. It is not uncommon for chemists working in companies to be
discouraged or prohibited, for proprietary reasons, from publishing
their work in journals or presenting at conferences. Therefore, patents
are a window onto chemistry research that would not otherwise be

visible. The importance of patents is also reflected by their increasing presence in chemistry research databases. For example, Reaxys, a chemical property and reaction database owned by Elsevier, contains about 4.6 million references of which 17% are patents. At the time of this writing, Chemical Abstracts, accessible via SciFinder® and STN, covers patents from 63 patent-issuing authorities.

3.3 WHAT IS A PATENT?

A patent is an exclusive right granted by a government to an inventor for a limited period of time, typically 20 years. In order to obtain a patent, an inventor must prove that his or her invention is novel, non-obvious and performs a useful (not necessarily practical) function. The inventor must also describe the invention in sufficient detail in order to enable a person having ordinary skill in the art to understand how the invention is made or used. In chemistry and related fields, a person with ordinary skill in the art might be defined as someone with an advanced degree. Secret ingredients and "black boxes" are not allowed in patent applications. Patent applicants who fail to disclose important information about a compound or process risk having their patent invalidated by the courts, as recently happened in the case of Pfizer's Canadian patent for Viagra®.[10] The requirement of full disclosure is a central component of the patent system: in exchange for full disclosure the inventor receives a 20-year monopoly; after the patent expires the technology becomes public domain. Once a patent has been granted, the owner has the exclusive right to prevent others from making, using, selling, offering for sale, or importing a product or a process, based on the patented invention, without the owner's permission. Monitoring and enforcement of patent rights is the responsibility of the patent owner.

In general, a patent may be granted for any new, useful machine, composition of matter, product, process, or improvement thereof. Subject matter that is generally not eligible for patent protection includes abstract ideas, artistic works, business methods, scientific theories, discoveries of substances as they naturally occur in the world, and inventions that are detrimental to public order, good morals, or public health. An example of the latter would be a method of producing and distributing a highly toxic substance, such as an anthrax letter bomb. Some countries restrict or prohibit the patenting of computer programs, diagnostic, therapeutic and surgical methods, chemical and pharmaceutical compounds (but not their methods of preparation), plants, and animals.

The types of patents of greatest interest to chemists cover new chemical compounds, mixtures, pharmaceuticals, processes, and methods of making a compound or improving the efficiency of a chemical synthesis. Some countries also allow patenting 3D atomic structures, structural databases (combinatorial libraries), and biological sequences.

Patent rights are territorial; there is no such thing as an international patent. Patents may only be enforced in the country or region in which they are issued. Patents may be granted by a national patent office, such as the U.S. Patent and Trademark Office, or a regional patent office for a group of countries, such as the European Patent Office. The cost of obtaining a patent varies widely from country to country. In the U.S., the current basic filing fee for a utility application is $390 ($195 for a business with fewer than 500 employees). In order to encourage applicants to file electronically, the USPTO offers a reduced basic utility application fee of $98 for small companies. Search and examination fees total about $870 ($435) and the issue fee is $1770 ($885). The minimum cost of obtaining a U.S. patent is about $3000 for large companies and $1500 for small firms and independent inventors. Of course, the actual cost may be several times higher, depending on the complexity of the application and whether or not the applicant hires a patent attorney to prosecute the application. The high cost of filing a patent application may discourage many independent inventors and start-up companies from seeking patent protection, especially for inventions in the early stages of development. In response, some countries allow inventors to file low-cost provisional patent applications. For example, the USPTO provisional patent application costs $250 ($125). U.S. provisional applications do not require as much information as a regular (non-provisional) application, are not examined, and must be converted into a regular application within 12 months.

In most countries, patent owners must pay maintenance or renewal fees at regularly intervals in order to maintain their rights. Failure to do so will result in the patent expiring before the end of its 20-year term. In the U.S., maintenance fees, due at 3.5, 7.5 and 11.5 years after issue, total $8860 ($4355). More than 50% of issued U.S. patents are maintained for 16 years, compared to 17 years for JPO patents and eight years for EPO patents.[11] Patent terms are non-renewable and once a patent has expired, even before the end of its 20-year term, it cannot be resurrected.

3.4 RELATIONSHIP OF PATENTS TO OTHER FORMS OF INTELLECTUAL PROPERTY

Patents are a form property known as intellectual property, or IP for short. Other types of IP include trade secrets, trademarks, and copyrights.

3.4.1 Trade Secrets

A trade secret is any piece of proprietary information that may be commercially valuable, including inventions that might be patentable. Unlike patents, trade secrets are not publically disclosed, never expire and require no application or maintenance fees. A well-kept trade secret can last indefinitely. The world's most famous, and perhaps oldest, trade secret is the formula for Coca-Cola, which was invented in the 1870s. However, trade secrets are vulnerable to reverse engineering and independent discovery. Companies can lose their trade secret advantage if their competitors learn how to make a product or process. Although trade secrets and proprietary processes are important in chemistry-related industries, companies often seek patent protection for their most valuable IP.

3.4.2 Trademarks

A trademark is a word, phrase or symbol that represents a product or service in the marketplace. Although they contain no chemical information, trademarks are very important in the chemical and pharmaceutical industries. Many chemical compounds, families of compounds and processes are known by their proprietary names rather than their scientific or common names. Examples include aspirin (acetylsalicylic acid), nylon (polyamides), Plexiglass® (methyl methacrylate) and Tylenol® (acetaminophen). Patent offices generally discourage the use of trademarks in patent applications. Indeed, the name of product may not be known at the time of filing, and many products are named long after a patent is obtained. In the case of pharmaceuticals and chemicals, many products are known by different trademarks in different countries. Chemical trademarks and trade names are indexed in many commercial chemical databases and some chemical search engines. However, trademarks are generally not indexed in public patent databases.

3.4.3 Copyrights

Copyright law protects "original works of authorship" that are fixed in a tangible form of expression. This includes literary works, computer programs, musical and dramatic works, pantomime and choreographic works, visual, graphic and sculptural works, motion pictures and other audiovisual works, sound recordings, and architectural works. In general, patent documents are not protected by copyright, although they may contain copyrighted material. For example, a chemical patent might include data and images that also appear in a scientific article written by the inventor.

3.5   TYPES OF PATENTS

There are many types of patents and patent documents. Some of the most common patents are described below. For more information regarding patent documents, see Sections 6 and 7.

3.5.1   Utility Patents

The most common and oldest type of patent is the patent of invention, also known in the U.S. as a utility patent. A utility patent protects new and useful inventions, *e.g.* devices, compounds, alloys, and processes. The term of a utility patent is twenty years from the date of filing of the application. In general, utility patent applications are published 18 months after the earliest filing date. This practice is a relatively recent development. For most of the 19th and 20th centuries, patent applications were kept confidential until a patent was granted. Beginning in the 1960s, countries began adopting early publication as a strategy to reduce workload and enhance public access to patent information. Although controversial among independent inventors and small companies, large companies generally welcomed early publication because it allowed them to make better informed decisions regarding research and development and patent applications. The USPTO began publishing applications in 2001. Figure 3.2 shows the front page of a recent U.S. chemical patent.

3.5.2   Design Patents (Industrial Designs)

The next most common type of patent is the design patent, also known as an industrial design in many countries. A design patent protects the ornamental design applied to a manufactured product. For example, the shape of a storage container or the design of protective lab shield may be protected by a design patent. Functional elements are not eligible for design patent protection, but an invention may be protected by both a patent and a design patent. In recent years, pharmaceutical companies have registered numerous design patents for shapes of pills, tablets, and capsules. Pfizer, Inc. patented sildenafil citrate, which it sells under the brand name Viagra®, and also registered a design patent for the uniquely shaped tablet.[12] Design patents have terms that are much shorter than regular patents. For example, the term of a design patent is

US008143432B2

(12) **United States Patent**	(10) **Patent No.:**	**US 8,143,432 B2**
Guillaume et al.	(45) **Date of Patent:**	**Mar. 27, 2012**

(54) **PROCESS FOR REGIOSELECTIVE MONO-TOSYLATION OF DIOLS**

(75) Inventors: **Michel Joseph Maurice André Guillaume, Berg (BE); Yolande Lydia Lang, Vosselaar (BE)**

(73) Assignee: **Janssen Pharmaceutica N.V.** (BE)

(*) Notice: Subject to any disclaimer, the term of this patent is extended or adjusted under 35 U.S.C. 154(b) by 318 days.

(21) Appl. No.: **12/515,407**

(22) PCT Filed: **Nov. 9, 2007**

(86) PCT No.: **PCT/EP2007/062109**

§ 371 (c)(1),
(2), (4) Date: **May 18, 2009**

(87) PCT Pub. No.: **WO2008/058902**

PCT Pub. Date: **May 22, 2008**

(65) **Prior Publication Data**

US 2010/0016623 A1   Jan. 21, 2010

(30) **Foreign Application Priority Data**

Nov. 17, 2006   (EP) 06124292

(51) **Int. Cl.**
C07C 303/04   (2006.01)
(52) **U.S. Cl.** **558/44**
(58) **Field of Classification Search** 558/44
See application file for complete search history.

(56) **References Cited**

FOREIGN PATENT DOCUMENTS

EP	0448413	9 1991
WO	WO 98 09942	3 1998

OTHER PUBLICATIONS

Boons et al., Synlett; No. 12; pp. 913-917 (1993).
David et al., *Tetrahedron* 1985. 41. 643.
Fasoli et al., *J. Mol. Cat. A* 2006. 244. 41.
Hua te al., Tetrahedron Letters; vol. 43; No. 48; pp. 8697-8700 (2002).

Martinelli et al., *J. Am. Chem. Soc.* 2002, 124. 3578.
Martinelli et al., Organic Letters. American Chemical Society; vol. 1; No. 3; pp. 447-450 (1999).
Murata et al., American Chemical Society; vol. 70; No. 6; pp. 2398-2401 (2005).
Seo et al., Journal of Organic Chemistry. vol. 72, No. 2, pp. 666-668, XP002469169 (2006).
Seo et al., "Supporting information" Journal of Organic Chemistry. [Online] XP002469170. American Chemical Society. Washington. DC. US Retrieved from the Internet: URL: http: pubs.acs.org subscribe journals joceah suppinfo jo061980u jo061980usi20061115_074917.pdf > [retrieved on Apr. 4, 2007] p. S5.
Shanzer, A. *Tet. Letters* 1980, 21. 221.
Tinsley, et al., Journal of the American Chemical Society; vol. 127; No. 31; pp. 10818-10819 (2005).
Tinsley, et al., "Supporting Information. Part I" Journal of the American Chemical Society, [Online] Jul. 13. 2005, pp. SI-SI8. XP002428306 American Chemical Society, Washington, DC, US, retrieved from the internet: URL: http: pubs.acs.org subscribe journals jacsat suppinfo ja0519861 ja0519861si20050610_013126. pdf.
Zarbin et al., Tetrahedron Letters; vol. 44; No. 36; pp. 6849-6851 (2003).

Primary Examiner — Joseph Kosack
(74) *Attorney, Agent, or Firm* — Yuriy P. Stercho

(57) **ABSTRACT**

The present invention concerns the use of dibutyl tin oxide for regioselective catalytic diol mono-tosylation at a concentration lower than 2 mol %. The present invention also concerns the use of a generic acetal compound of Formula (Ic) in a catalytic process for regioselective diol mono-tosylation, wherein Y is selected from the group of C_{1-6} alkyl, phenyl and benzyl. The concentration of the generic acetal compound of Formula (Ic) is less than about 2 mol %, preferably ranges between about 2 mol % and about 0.0005 mol %, preferably ranges between about 0.1 mol % and about 0.005 mol %.

(Ic)

3 Claims, No Drawings

Figure 3.2   Process for Regioselective Mono-tosylation of Diols, US 8,143,432 B2.

14 years in the U.S. and 10 years in Canada. Design patents are of little interest to chemists because they contain no useful chemical information. However, many pieces of equipment found in a chemistry lab may protected by design patents. Figure 3.3 is the front page of a recent US design patent for a reagent cartridge.

3.5.3   Plant Patents

Since 1930, inventors with green thumbs have been able to obtain patent protection in the U.S. for distinct and new varieties of asexually reproducible plants (that is, through graftings, cuttings, tissue cultures, *etc.*). This includes flowers, fruit trees, shrubs, vegetables, grass, mushrooms, and the like. Algae and macro fungi are eligible for plant patent protection, but bacteria are not. Like utility patents, plant patents are granted for twenty years. Approximately 23 000 plant patents have been granted to date, the majority to residents of the U.S., Denmark, Germany, Great Britain, and The Netherlands. Protection for sexually reproducing plant varieties may be obtained through UPOV, the International Union for the Protection of New Varieties of Plants (www.upov.int).

3.5.4   Utility Models

Utility models protect minor improvements that are not eligible for conventional patent protection. Some countries register utility models without examination. Like design patents, utility models have terms that are shorter than standard patents, typically 5 to 10 years. Countries that offer utility model protection include Australia, China, Germany, Japan, Korea, and Mexico. Chemical compounds and processes of making them are not eligible for utility model protection, so these documents are generally of little interest to chemists. However, lab instruments and apparatus could be eligible for utility model protection.

3.5.5   Supplementary Protection Certificates

Although the term of a granted patent cannot be renewed or extended, some countries allow term extensions for certain types of patents, generally those covering pharmaceutical and agrochemical inventions. SPCs and patent term extensions are granted in order to compensate patent owners for the time it takes to obtain permission from government health and public safety agencies to market a new drug or agricultural chemical. In the U.S. and Japan, a patent term may be extended by

US00D645973S

(12) **United States Design Patent**　(10) **Patent No.:**　　**US D645,973 S**

Hoenes　　　　　　　　　　　　　　　(45) **Date of Patent:**　　** Sep. 27, 2011

(54) **REAGENT CARTRIDGE FOR HOLDING REAGENTS IN A CHEMISTRY ANALYZER**

(75) Inventor: **Erik Robert Hoenes**, Laguna Niguel, CA (US)

(73) Assignee: **Beckman Coulter, Inc.**, Brea, CA (US)

(**) Term: **14 Years**

(21) Appl. No.: **29/379,633**

(22) Filed: **Nov. 22, 2010**

(51) LOC (9) Cl. **24-01**
(52) U.S. Cl. **D24/224**
(58) Field of Classification Search D24/216 232, D24/130, 169; D10/81; 422/500, 547, 549, 422/556, 557, 558; 435/287.1, 288.1
See application file for complete search history.

(56) **References Cited**

U.S. PATENT DOCUMENTS

D149,313 S	4	1948	Shapiro
D159,762 S	8	1950	Touhey
3,295,710 A	1	1967	Marchant
3,554,705 A	1	1971	Johnston et al.
3,582,286 A	6	1971	Hamilton
3,713,985 A	1	1973	Astle
3,788,815 A	1	1974	Rohrbagh
3,883,815 A	1	1974	Hoskins et al.
3,905,772 A	9	1975	Hartnett et al.
3,907,505 A	9	1975	Beall et al.
D237,655 S	11	1975	Engelsher
3,964,831 A	6	1976	Frank
3,964,867 A	6	1976	Berry
3,994,594 A	11	1976	Sandrock
4,043,678 A	8	1977	Farrell et al.
4,083,638 A	4	1978	Sandrock
D250,834 S	1	1979	Ruppert
4,178,152 A	12	1979	Nunogaki
4,200,613 A	4	1980	Alfrey et al.
4,234,540 A	11	1980	Ginsberg et al.
4,251,159 A	2	1981	White

4,274,885 A	6	1981	Swartout
4,287,155 A	9	1981	Tersteeg et al.
D261,552 S	10	1981	Boroda
4,344,768 A	8	1982	Parker et al.
4,346,056 A	8	1982	Sakurada
4,360,360 A	11	1982	Chiknas
4,371,498 A	2	1983	Scordato et al.
4,387,164 A	6	1983	Hevey et al.
4,391,780 A	7	1983	Boris
D276,264 S	11	1984	Boris
D276,265 S	11	1984	Boris
D276,266 S	11	1984	Boris

(Continued)

FOREIGN PATENT DOCUMENTS

EP　　　136125　　　4 1985

(Continued)

Primary Examiner — Anhdao Doan
(74) *Attorney, Agent, or Firm* — Merchant & Gould PC

(57) **CLAIM**
The ornamental design for a reagent cartridge for holding reagents in a chemistry analyzer, as shown and described.

DESCRIPTION

FIG. 1 is a top perspective view of a reagent cartridge for holding reagents in a chemistry analyzer showing my new design;
FIG. 2 is a front elevation view thereof;
FIG. 3 is a rear elevation view thereof;
FIG. 4 is a left side elevation view thereof;
FIG. 5 is a right side elevation view thereof;
FIG. 6 is a top plan view thereof;
FIG. 7 is a bottom plan view thereof; and,
FIG. 8 is a bottom perspective view thereof.
The broken lines depict portions of the reagent cartridge for holding reagents in a chemistry analyzer that form no part of the claimed design.

1 Claim, 8 Drawing Sheets

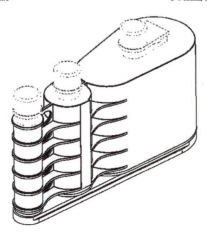

Figure 3.3 Reagent Cartridge Holder for Holding Reagents in a Chemistry Analyzer, US D645,973.

simply extending the term of the existing patent. In Europe, however, patent owners must obtain a separate Supplementary Protection Certificate (SPC) in order to extend a patent's term.

3.5.6   Patent Gazettes, Journals, Abstracts, and Indexes

Prior to the rise of the internet in the 1990s as a means of disseminating timely patent information, patent offices published abstracts of issued patents in the form of printed gazettes and journals. One of the earliest patent abstracts was the weekly *Official Gazette of the U.S. Patent Office*, which was first published in 1872. In addition, many patent offices published annual patent indexes which were useful for locating patents by inventor name, assignee, or subject matter. Today, most printed patent gazettes and indexes have been superseded by electronic versions or online patent databases.

3.6   PATENT PROCESS FROM INVENTION TO PUBLIC DOMAIN

3.6.1   Patent Process

The process of patenting an invention begins in the mind of the inventor. Anyone can be an inventor. The process of inventing is not limited by gender, age, nationality, citizenship, profession, education, physical, or mental state. Even deceased inventors can obtain patents through their estates. A patent application must include at least one named inventor who is a real person (pseudonyms are not permitted). Companies and organizations cannot be inventors, although they may apply for a patent on behalf of an inventor.

There is no limit to the number of joint inventors who can apply for a patent, but all must be able to prove that they made a meaningful contribution to the invention. In large academic and industrial research labs, where teams of researchers work together, joint inventorship is fairly common. Of course, only the most complex inventions could legitimately have multiple co-inventors, and the vast majority of patents have fewer than ten inventors. Disputes involving joint inventorship are not uncommon and can be difficult to resolve, especially in settings such as universities, where collegiality is the norm and there is frequent turnover of graduate students, post-doctoral fellows and visiting professors.

In most countries, an inventor may file an application without the assistance of a patent attorney or patent agent. This is known as filing a *pro se* application, *pro se* meaning to represent oneself in a legal proceeding. However, because patent law is so complex and evolving constantly, inventors are well advised to seek the assistance of a licensed patent professional. The vast majority of patent applications are prosecuted by patent agents. An individual who wishes to become a registered patent agent must first pass a qualifying exam administered by the patent office. A patent agent is someone with a technical background, such as an undergraduate degree in chemistry, who has passed the qualifying exam. A patent attorney is someone who is licensed to practice before the patent office and by the local or national attorney licensing authority.

Filing requirements vary from country to country. In general, a patent application must contain the following documentation.

3.6.1.1 Inventor and Legal Representative Information. A patent application must contain a formal petition from the inventor. The petition is a statement by the inventor declaring that he or she is the original and sole (or joint if more than one inventor is listed) inventor of the invention disclosed in the application. The inventor must also provide his or her residence, citizenship, and address (or the patent agent's name and address). If the inventor has hired a patent agent to prosecute the application, a letter granting power of attorney must be included.

3.6.1.2 Title and Abstract of the Invention. The title is a short technical description of the invention generally limited by patent office rules to a few hundred characters. The abstract of the invention (also called the abstract of disclosure) is a brief summary of the contents of the written description, including the most novel and important characteristics of the invention. For chemical patents, patent office rules may require that the abstract include a general description of the nature of the compound or composition and its use. For example, "the disclosed invention describes a novel class of macrocyclic compounds which are useful as luminescent markers". For processes, the type of reaction and reagents should be described. Patent examiners may request changes to the title and abstract to conform with these guidelines.

3.6.1.3 Written Description of the Invention (the Specification). The written description, also known as the specification, includes a detailed explanation of the drawings, the background of the invention, which may include a discussion of the prior art, and a description of how the invention is made or used. The written description must provide sufficient detail so that anyone skilled in the art can comprehend and practice the invention.

3.6.1.4 Drawings. The drawings are black and white diagrams that show the technical details of the invention. Drawings are not required for all inventions. For example, inventions relating to processes or methods generally do not require drawings. A model of the invention or sample of the chemical compound is also not required.

3.6.1.5 Claims. The claims appear at the end of the written description, generally in the form of numbered paragraphs or sentences. The claims define the legal scope of protection and are the most important part of the patent application. Below is an example of a claim from a U.S. patent, US 8,143,432 B2.
  "The invention claimed is:

1. Process for the catalytic mono-tosylation of a diol, comprising a step wherein a compound comprising a diol moiety of Forumla (Ia) is tosylated into a compound comprising a tosylated diol moiety of Formula (Ib) using a compound of Formula (Ic) Where inY is selected from the group C1-6, alkyl, phenyl, and benzyl, in a concentration between 2 mol % and 0.0005 mol %."[13]

An application must have at minimum one claim; there is no limit to the number of claims that may be stated.

3.6.1.6 Known Prior Art. Applicants may be required to disclose all known prior art in their applications. Prior art is defined as all the relevant technical information that is available to the public in any form anywhere in the world before a certain date that might be relevant to the patentability of an invention. Prior art includes patent documents and non-patent literature, *e.g.,* journal articles, theses, books, presentations, websites, and conference papers.
  If the application is deemed complete by the receiving patent office, an official file will be opened and assigned to a patent examiner with expertise in the appropriate technology. Depending on the country, the applicant may have the option of deferring examination for several years. For example, in Canada, applicants may defer examination up to five years. The U.S. does not allow applicants to defer examination. The application process may take two years or longer, depending on the complexity of the disclosed invention and the backlog of pending applications in the office of filing. At the USPTO, the current average pendency (the time from filing to the time a patent issues or the application abandoned) is about 33.2 months.[14] Pendency rates vary significantly by technology. For example, the USPTO pendency for biotechnology and organic chemistry applications is 33.6 months; for chemical and materials engineering it is 34.6 months. However, the

U.S. recently introduced a prioritized examination option for utility and plant patent applications that allows applicants, for a fee, to accelerate the examination process. The goal of the program is to provide a final decision on a patent application within 12 months.

In order to be granted a patent, an applicant must demonstrate that his or her claimed invention meets the following criteria. First, the invention must consist of patentable subject matter as defined by under local patent law. Second, the invention must be novel (new), not part of the prior art. Third, the invention must involve an inventive step and be non-obvious to the average person skilled in the art. The invention must also be useful and capable of industrial application. Finally, the invention must be fully disclosed in the patent application.

3.6.2 Patenting Abroad

Inventors who wish to obtain patent protection for their invention in more than one country have two options. Under the provisions of the Paris Convention for the Protection of Industrial Property, they may file separate applications in each country they wish to obtain patent protection, a process known as the national route. Alternatively, inventors may file a single international patent application under the Patent Cooperation Treaty (PCT) system. The PCT system was created in the 1970s in order to facilitate the process of obtaining patent protection in more than one country. Membership is open to countries that are parties to the Paris Convention. There are now 146 contracting parties in the PCT system, the most recent additions being Panama and Brunei. The PCT is administered by the World Intellectual Property Organization (WIPO), an agency of the United Nations based in Geneva, Switzerland. Applicants may file their international patent application directly with the WIPO or *via* their national patent office.

Applications filed under the PCT are not examined by the WIPO and do not become patents. Instead, a PCT application becomes the basis for national patent applications, which are examined by the patent offices in PCT member countries designated by the applicant.

Most national patent laws require foreign applicants to be represented by a registered patent agent who is a resident of the country. Rosters of registered patent attorneys and agents are maintained by a number of patent offices, including the USTPO and Canadian Intellectual Property Office.

3.6.3 Patent Families

When an inventor files separate patent applications that describe the same invention in more than one country a patent family is born. A patent family is a group of published patent documents that disclose the same invention and are linked by one or more common priority numbers. A priority number is the application serial number for the earliest application, including U.S. provisional applications, in a group of documents. A patent family may have as few as two members up to several hundred. The patent family for Warner-Lambert's "Polymer Compositions Containing Destructurized Starch" (US 5,095,054) has at least 270 family members. Patent family information is useful for determining the scope of international patent protection for a specific patented product or process. It can also be helpful for locating a translation of a patent document. One of the best public patent databases for searching patent family information is the EPO's Espacenet system.

There are several ways of defining a patent family. For example, the EPO defines a patent family in Espacenet as all documents having exactly the same priority or combination of priority numbers. This is known as a simple patent family. Figure 3.4, below, shows the simple patent family for a U.S. patent.

A complex patent family is a group of documents disclosing the same invention that have one priority application number in common with all other members of the family.

The extended patent family, which is used in INPADOC, is defined as a group of published patent documents, each having at least one priority number in common with at least one other document in the group. This definition also takes into account national and international application

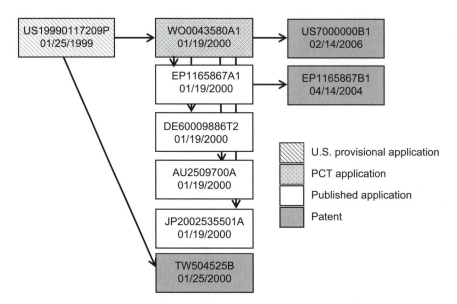

Figure 3.4   Patent Family Tree for US 7,000,000 B1.

Table 3.2 Simple, Complex, and Extended Patent Families.

D1	Priority P1		
D2	Priority P1		
D3	Priority P1	Priority P2	
D4		Priority P2	Priority P3
D5		Priority P2	Priority P3
D6			Priority P3

numbers and can result in rather large groups of documents. CAS applies the simple patent family definition to include patents in the "basic" patent record in CAplus/SciFinder. However, CAS includes extended patent family information for some patents that have more complex family relationships. This can lead to multiple records linked by common priority numbers. The PatBase system also uses the extended patent family definition.

In the Table 3.2, documents D1 and D2 comprise a simple patent family because they share the exact same priority number (P1). D3 is not a member of this family because it has two priority numbers, P1 and P2. Documents D1, D2 and D3 make up complex patent family P1 because they share a common priority number (P1). Document groups D3-D4-D5 and D4-D5-D6 are also complex patent families. Under the extended family definition, all six documents (D1-D6) would comprise a patent family because they share at least one priority number in common with at least one other member of the family.

An artificial patent family is a group of equivalent patent documents published by different offices that do not share common priority numbers but disclose the same invention. This type of family arises when an inventor files separate applications in two or more countries that do not reference one another. An example of an artificial patent family is US 1,703,788 and CA 263,349, which disclose Canadian inventor Arthur Sicard's snow removing machine (snow blower).

A national patent family is a group of related documents published by the same office at least two of which are distinct from one another. This could include the published applications and issued patents originating from a single application.

3.6.4 Legal Status and Ownership (Assignment)

After a patent has been granted and during the course of its 20-year life, a number of events may occur that could affect its legal status and ownership. Perhaps the most common event is the expiration of the patent due to the owner's failure to pay maintenance fees. Transfer of ownership (re-assignment) of granted patents and pending applications is also common. A patent printed with typographical error or some other minor defect can be corrected by issuing a new document or in the case of the U.S., appending a certificate of correction to the original document. Historically, changes in legal status and ownership were difficult to determine because few patent offices made this information available on the internet. In the past ten years, however, patent offices have made great improvements in this area. For U.S. patents, changes of ownership from 1980 to the present can be found in the USPTO's patent assignment database (http://assignments.uspto.gov).

3.7 PATENT DOCUMENTS

Patent documents include all forms of published patent applications, issued patents, and supplemental documents such as U.S. certificates of correction. Historically speaking, patent documents varied widely in format from country to country. During the last fifty years, however, patent offices have standardized patent document formats to such a degree that there is very little difference between countries.

3.7.1 INID Codes

INID (Internationally Agreed Numbers for the Identification of (bibliographic) Data) codes are used by patent offices worldwide to identify specific data elements on the front page of a patent document. They are defined in WIPO Standard ST.9.[15] The purpose of INID codes is to help patent searchers navigate patent documents written in languages they do understand. See Table 3.3 for a list of the most common types of INID codes.

3.7.2 Patent Numbers

Patent offices employ a variety of numbering systems to identify, track, and link patent documents. In some cases, the number of different systems used within a patent office can be bewildering to an inexperienced patent searcher. For example, a U.S. patent may have four or more unique numbers associated with it. These include a provisional application number, an application number, a publication number and an issue number. Reissue patents and other supplemental patent documents may be assigned their own unique numbers.

The Application Number is the number assigned to the application when it is received by the patent office. The Publication Number is assigned when the application is published at 18 months. The Issue Number is the number assigned when the patent is granted. The Priority Number is the number assigned to the earliest patent application

Table 3.3 Selected INID Codes.

Code	Description
(11)	Patent number
(12)	Type of document
(21)	Application number
(22)	Filing date
(43)	Issue date
(45)	Publication date
(51)	International Patent Classification
(54)	Title
(57)	Abstract
(71)	Applicant
(72)	Inventor
(73)	Assignee

(including provisional applications) in a group of related documents called a patent family. Patent families are generally defined as one or more applications having a common priority number or numbers.

Although there has been much effort over the last several decades to standardize patent number formats, every patent office has its own unique numbering systems, some of which are based on local practices and customs. A well-known example is the Japan Patent Office's numbering system, based on the Japanese Year of the Emperor, which was in use until 2000 when the JPO adopted the western year format. Depending on the patent office, a patent application may be assigned several unique identification numbers as it moves through the examination process. Patent offices may also use different types of numbering systems for different types of patent documents. For example, the USPTO has separate numbering systems for published applications (US-A), issued patents (US-B), design patents (US-S), reissue patents (US-E), plant patents (US-P) and statutory invention registrations (US-H).

The current standard format for a patent publication number is shown below.

CC 1,234,567 KK

CC is the country code, a two-letter code representing the country or patent office of origin as defined in WIPO Standard ST.3.[16] KK is the kind code, a one or two letter code indicating the type of patent document as defined in WIPO Standard ST.16.[17] Selected country and kind codes are shown in Tables 3.4 and 3.5, respectively.

3.7.3 Front Page, Drawings, Specification, and Claims

The standard patent document includes a front page containing bibliographic information about the patent, an abstract, list of prior art

Table 3.4   Selected Country Codes.

Code	Country/Office
AU	Australia
CA	Canada
BR	Brazil
CN	China
DE	Germany
EP	European Patent Office
FR	France
GB	United Kingdom
IT	Italy
JP	Japan
KR	Korea, Republic of
MX	Mexico
RU	Russia
US	United States
WO	PCT application

Table 3.5   Selected Kind Codes.

Code	Document Type
A	First publication (patent application)
B	Second publication (patent)
C	Third publication (patent)
E	Reissue patent document (US)
H	Statutory invention registration (US)
P	Plant patent document (US)
S	Design patent document (US)
T	Translated application
U	Utility model application
Y	Utility model registration

references, and representative drawing; the front page is followed by the drawing pages and written specification. The drawing pages feature black and white technical drawings (and sometimes photographs) of the invention or process. For chemistry patents, drawings might consist of one or more chemical structures.

3.8   PATENT SEARCHING

3.8.1   Types of Searches

3.8.2   Searching Inventor and Company Names

Few inventions are known by their patents. How many people would recognize US 7,345,671, "Method and Apparatus for Use of Rotation

User Inputs", as Steve Jobs' design for the iPod user interface? Great inventions are known by the names of their inventors or the companies that bring them to market.

While academic chemists publish most of their research in journal articles, patents are often the predominant type of publication produced by industrial chemists. For example, Du Pont research chemist Stephanie L. Kwolek published approximately 30 articles and 17 U.S. patents during her forty-year career. Her patents have been cited in 382 journal articles indexed in Science Citation Index and 267 U.S. patents in the USPTO patent database. In 1995 Kwolek was inducted into the National Inventors Hall of Fame for her invention of Kevlar®. Italian chemist and Nobel prize (1963) laureate Guilio Natta published some 471 journal articles and 238 patents during his lifetime.

Retrieving patent documents by searching inventor names is relatively straightforward in most public and commercial patent databases. However, there are several factors that patent researchers should take into account in order to improve their search results.

Variations in the spelling of inventor names can be problematic.[18] For example, Michael Lazaridis, the Canadian inventor of the BlackBerry® smartphone and co-founder of Research in Motion, is listed as Mihal Lazaridis on most of his patent applications. Lazaridis' parents emigrated to Canada from Greece; "Mihal" is the Greek spelling of "Michael". Likewise, the transliteration of non-English letters with accent marks can give rise to many name variations. For example, Helmut Möhwald appears in numerous forms on patent documents and in database records, including Helmut Moewald, Helmut Moehwald, Helmuth Moehwald, Helmut Mohwald, Helmuth Mohwald, Helmut Möhwald and Helmuth Möhwald (See Table 3.6). Misspelled names, *e.g.* Hellmuth Mohwald, can further complicate name searches.

The transliteration of western names into Asian languages and *vice versa* can also create spelling variations that are extremely challenging to find. For example, Normand Croteau might appear on a

Table 3.6 Inventor Name Variations in Chemistry/Patent Databases (May 30, 2012).

	SciFinder (CAplus)	*Espacenet*	*PatentScope*	*USPTO*
Moehwald, Helmut	35	51	48	10
Moehwald, Helmuth	24	36	23	1
Moewald, Helmut	1	1	0	0
Moewald, Helmuth	0	0	0	0
Mohwald, Helmut	13	28	7	24
Mohwald, Helmuth	5	24	2	19

Korean patent as Normang Croto and Mark Loring Smith appears as Maaku Rooringu Sumisu and Valentina Riveros-Rojas as Barenteina Riberosuurojasu on JP documents. The Japanese surname "Namba" is often spelled Nanba on US patent documents. This problem is likely to grow with the increasing numbers of patent applications filed in Asian countries, especially China and Korea.

Some inventors use their initials or nicknames on patent applications. For example, Nobel Laureate Dr Harold Kroto is listed as Harry Kroto on a WO application (WO02088024 A1) for coated nanotubes. Dr Harold Steenbock, a professor of biochemistry at the University of Wisconsin who in the 1920s invented a method of increasing the vitamin D content of foods, also goes by Harry on his patents. Dr KR Sindhar, co-founder and CEO of Bloom Energy, a manufacturer of fuel cells, is listed simply as "K. R. Sindhar" on his patent applications. Abbreviated first names are common on nineteenth-century patents. For example, Richard may appear as Richd. and Joseph as Jos. When searching for a woman inventor's patents, searchers must take into account any changes in the last name due to marriage or divorce.

Some patent search systems take advantage of patent family linkages to overcome the problem of spelling variations and errors in names. For example, a name search in Espacenet will retrieve all patent family members, increasing the likelihood of retrieving documents with name variations.

Company and organization names usually appear in the applicant or assignee fields on patent documents and database records. As with inventor names, variations in spelling can make searching challenging. Some names are fairly easy to deduce. IBM's patents can be retrieved using the company's full name, International Business Machines. Canadian chemical company C-I-L, Inc., manufacturer of paints, pesticides, fertilizers, and explosives, was formerly known as Canadian Industries Limited.

Other company names have long and complicated histories. For example, the American chemical giant Du Pont has been known by several names since it was founded as a gunpowder manufacturer in Delaware in the early 1800s. For much of the past century it has operated as E.I. Du Pont de Nemours & Co; searching this name in Espacenet will retrieve more than 100 000 patent documents. Du Pont also has many subsidiaries around the world that file (or have filed) patent applications under a local name. Examples include Du Pont of Canada and Du Pont de Nemours (Deutchland) GbmH.

Inventors may not change their names often, but it is not uncommon for companies to do so. For example, 3M was formerly known as

Table 3.7 Company Name Variations in Chemistry and Patent Databases (June 30, 2012).

Assignee Name	SciFinder (CAplus)	Espacenet (US only)	USPTO (1976–)
Minnesota Mining & Manufacturing	7,127	9,514	11
3M	7,282	8,972	6,711
Pittsburgh Plate Glass	1,646	2,394	0
PPG Industries	3,987	5,125	3,835
Queen's University	8	0	31
Queen's University at Kingston	239	0	238
Univ Kingston	0	323	0

Minnesota Mining and Manufacturing. The American chemical company PPG Industries was once called Pittsburgh Plate Glass. Sometimes one character or space in a company's name can make a huge difference in retrieving its patents. In 2008, Dutch chemical company Akzo Nobel rebranded itself, changing its public name to AkzoNobel. However, its patents can still be found under its former two-word name.

Name changes often arise from mergers and acquisitions. When Lyondell Chemical Co. and Basell Polyolefine merged in 2007, the new company was named LyondellBasell Industries (LBI). In order to get a complete picture of LBI's patent portfolio, one would have to search for patents under the current name, plus the names of the two predecessor companies.

Transliteration problems can also arising with assignee names. The Wellcome Foundation appears as UERUKAMU FUAUNDEESHON on some Japanese patent documents and Sparton Resources, a mineral and energy company based in Toronto, is rendered Beijing Sipadum Mineral Resources on Chinese documents.

Some commercial database vendors normalize and index names as an aid to searchers. Public and commercial databases may also use special codes or abbreviations to identify companies. For example, in Espacenet, patents assigned to Queen's University, located in Kingston, Ontario, Canada are indexed under "Univ Kingston" (See Table 3.7).

3.8.3 Searching Cited References

Patent documents often include cited references to patents and non-patent literature (NPL), *i.e.,* prior art, related to the technology disclosed in the application. NPL includes journal articles, conference papers, books, theses, dissertations, standards, and gray literature. Electronic sources such as databases and websites also may be cited.

Inventors are required to disclose all known prior art when they file their application, and during the examination process if new information is discovered. Patent examiners may also identify prior art during the examination process. Inventor-supplied references are provided on an Information Disclosure Statement (IDS) that is filed with the application or included in the "Background of the Invention" that forms part of the written specification. References identified by the patent examiner are printed on the search report accompanying the application (EP-A, WO-A) or on the issued patent (EP-B, JP, US-B). Cited references were first included on US patents in 1947. However, references do not appear on US published applications (US-A). On U.S.-issued patents, references cited by the examiner are marked with an asterisk.

The ability to search cited references is very useful for locating other relevant patents and NPL. Many patents contain dozens or even hundreds of references, in effect providing a historical review of a technological problem or technology. References often reveal the surprising historical origins of modern inventions. For example, the patent (US 4,253,132) for the TASER™ stun gun cites an 1852 patent (US 8,843) for an electrified whale harpoon! This is especially useful for discovering early patents that are not fully indexed in a database.

Searching cited references is also useful for identifying important patents and inventors. For example, Nobel Laureate Dr Kary Mullis' 26 U.S. patents have been cited more than 7000 times; his 1987 patent (US 4,683,202) for a process for amplifying nucleic acid sequences has been cited in 2521 U.S. patents and more than 3300 patents worldwide.

It is possible to search cited patents in a number of patent and non-patent literature databases. Public patent databases with this capability include the USPTO, Espacenet, and DEPATISnet. In Web of Science (Science Citation Index), patents cited in journal articles are searchable by inventor name and number.

The Common Citation Document (CCD) system (http://www.epo.org/searching/free/citation.html) is a new citation search tool created by the European Patent Office, Japan Patent Office and USPTO. It combines prior art cited in EP, JP and US patent documents, which can be retrieved by patent number, into a single easy-to-read page.

3.8.4   Patent Classification Searching (Subject Searching)

A patent classification system is a rules-based scheme designed with the purpose of facilitating the organization, indexing, storage, and retrieval of patent documents. In the 1800s, patent offices devised a variety of patent classification systems in order to organize their own national patent document collections, foreign patent documents, and non-patent literature. By the mid-1950s, the rising expense of maintaining separate national classification systems prompted several patent offices in Europe to create a shared international classification system. This initiative evolved during the 1960s into the International Patent Classification (IPC) system, which was formally adopted by 15 countries under the Strasbourg Agreement of 1971. Since the 1970s, many patent offices have abandoned their national patent classification systems in favor of the IPC. The major patent classification systems in use in 2012 include the International Patent Classification (IPC), European Classification (ECLA), Japan Patent Office F-term and File Index systems, and the U.S. Patent Classification (USPC). The USPTO and the EPO are currently working on a Cooperative Patent Classification (CPC) system that will replace the USPC and ECLA beginning in 2013. The Derwent Patent Classification is a proprietary patent classification system owned by Thomson Reuters which is used in the Derwent World Patent Index database. This section will focus primarily on the classification of chemical subject matter in the USPC and IPC systems.

3.8.4.1 U.S. Patent Classification System (USPC). The USPC in its present form was established in the late 1890s. Over the last one hundred years it has grown from approximately 200 classes to more than 450 classes.[19] It has undergone numerous reorganizations and is continuously revised in order to accommodate new technological innovations. The USPTO updates classification data in its internal and public databases approximately bimonthly. However, it does not update the classifications printed on patent documents. The USPC is one of the last national patent classification systems in use today but will not be employed for much longer. Beginning in 2013, it will be phased out and replaced by the Cooperative Patent Classification (CPC) system. However, the transition is likely to take several years.

The USPC currently consists of 431 utility classes numbered 2 through 850, six cross-reference classes numbered 901 through 976, 32 design classes numbered D1 through D99, and one class designated PLT for plant patents. Class 1 is reserved for administrative purposes. Each class is accompanied by a set of definitions that provides detailed descriptions about the technological subject matter covered in the class and "see also" references to related classes. The *Index to the USPC*[19] is an alphabetical list of technical and common terms and trade names and their corresponding classes or subclasses whose purpose is to help patent searchers quickly locate appropriate the appropriate class and subclass for a given invention. For example, Figure 3.5 shows the *USPC Index*

Boron and Compounds
Acid and acid anhydride ....   423 / 277
Binary compounds ....   423 / 276+
Borates   423
Carbide ....   423 / 291
Carbocyclic or acyclic compounds ....   568 / 1+
Cleaning or detergents ....   510 / 345
Compositions containing ....   510 / 345
Cleaning or detergents ....   510 / 465
Compositions containing ....   510 / 465
Elemental ....   423 / 298
Esters containing ....   423 / 277
Explosive or thermic ....   149 / 22
Composition containing ....   149 / 22
Fuels containing ....   44 / 314+
Nuclear fuel containing ....   252 / 636+

Figure 3.5   USPC Index Listing for Boron and Boron Containing Compounds.

CLASS 560, ORGANIC COMPOUNDS
1   . Carboxylic acid esters
129   . . Acyclic acid moiety
130   . . . Esterified phenolic hydroxy
143   . . . . Salicylic acid or functional derivative

Figure 3.6   USPC Schedule (Collapsed) for Salicylic Acid.

entry for boron and boron containing compounds. The + symbol following a subclass indicates that the entry includes that subclass and all subclasses indented below.

Although the USPC system is based on hierarchical principles, it has a relatively flat structure. The top level consists of about 470 classes, while the second level consists of approximately 150 000 subclasses. A USPC classification, therefore, consists of two parts, a class followed by a subclass. For example, 560/143 is the classification for salicylic acid (aspirin). The hierarchical level among subclasses is indicated by dots and the placement of subclasses within a schedule. (See Figure 3.6) This dual-level structure is also found in design and plant classes. For example, D24/101 is the design code for a pharmaceutical tablet or caplet. The number of subclasses within a class ranges from as few as a dozen to a thousand or more. Subclasses that become too large due to the growing number of patent documents are divided into multiple subclasses. Classes that become too large may also be divided. For example, Classes 260, 518, 520–529, 530–570 are considered one integral class for organic compounds.

There is no fixed relationship between class or subclass numbers and technological subject matter. It is not uncommon to find proximate classes covering very different subject matter.

USPC classes are organized into four subject matter groups:

- Group I. Chemical and Related Arts.
- Group II. Communications, Radiant Energy, Weapons, Electrical, and Computer Arts.
- Group III. Body Treatment and Care, Heating and Cooling, Material Handling and Treatment, Mechanical Manufacturing, Mechanical Power, Static, and Related Arts.
- Group IV. Designs.

Group I includes approximately 95 classes relating to chemistry, chemical processes, and chemical equipment. This represents about 22% of the USPC. Group I also includes classes relating to ceramics, metallurgy, biotechnology, and nanotechnology.

The largest chemistry-related USPC classes (in terms of the number of classified patent documents) are Class 260, Chemistry of Organic Compounds; Class 423, Chemistry of Inorganic Compounds; Classes 424 and 514, Drug, Bio-Affecting and Body Treating Compositions; Classes 520–528, Synthetic Resins or Natural Rubbers; Classes 532–570, Organic Compounds; and Class 585, Hydrocarbons.

Class 260 is the generic class for organic compounds (see Figure 3.7). It was at one time the largest class in the USPC. Organic compounds are defined as compounds meeting at least one of the following criteria:

(1) two carbon atoms bonded to each other;
(2) one carbon atom bonded to at least one hydrogen atom or halogen atom; or
(3) one carbon atom bonded to at least one nitrogen atom by a single or double bond.

The exceptions to this rule are: hydrocyanic acid, cyanogen, isocyanic acid, cyanamide, cyanogen halides, isothiocyanic acid, fulminic acid, and metal carbides, which are classified as inorganic compounds.

3.8.4.2 International Patent Classification (IPC). The IPC is currently used by more than 100 countries including the U.S. and Canada. It is administered by the WIPO and various committees and working groups whose members are from countries who are parties to the Strasbourg Agreement of 1971.

The IPC is a hierarchical classification consisting of five levels.[20] The top level consists of eight sections, which are further subdivided into

Class 260	Chemistry of Organic Compounds
. Class 518	Fischer-Tropsch Processes
. Class 520	Synthetic Resins & Natural Rubbers
. . Class 521	Ion-Exchange Polymers, Cellular Products, Waste Polymer Recovery
. . Class 522	Wave Energy Polymer Chemistry
. . Class 523	Synthetic Resin Compositions with Nonreactant Material
. . . Class 524	Class 523 continued
. . Class 525	Chemically Treated Synthetic Resins, Compositions of Plural Synthetic Resins
. . Class 526	Miscellaneous Processes, Synthetic Resins from Only Ethylenically Unsaturated Monomers
. . Class 527	Synthetic Resins from Specified Natural Sources
. . Class 528	Synthetic Resins from Plant Material of Unknown Constitution or Specified Reactant
. Class 530	Natural Resins, Peptides, Proteins, Lignins
. Class 532	Class 260 continued
. . Class 534	Radioactive or Rare Earth Metal Compounds, Azo and Diazo Compounds
. . Class 536	Carbohydrates
. . Class 540	Heterocyclic Carbon Compounds
. . . Class 544	Six-Membered Nitrogen Hetero Rings with Two or More Hetero Atoms
. . . Class 546	Six-Membered Hetero Rings with One Ring Nitrogen
. . . Class 548	Five-, Four- or Three-Membered Nitrogen Hetero Rings
. . . Class 549	Oxygen or Sulfur Hetero Rings
. . Class 552	Azides, Triphenylmethanes, Quinones, Hydroquinones, Steroids
. . Class 554	Fats, Fatty Derivatives
. . Class 556	Heavy Metal, Aluminum or Silicon Compounds
. . Class 558	Esters
. . Class 560	Esters continued
. . Class 562	Acids, Acid Halides, Acid Anhydrides, Selenium & Tellurium Compounds
. . Class 564	Amino Nitrogen Compounds
. . Class 568	Boron, Phosphorus, Sulfur, or Oxygen Compounds
. . Class 570	Halogen Compounds
. Class 585	Hydrocarbons

Figure 3.7   Hierarchy of USPC Class 260 and Its Integral Classes.

classes, subclasses, main groups, and sub-groups. The top level sections are:

A. Human Necessities;

B. Performing Operations; Transporting;

C. Chemistry; Metallurgy;

D. Textiles; Paper;

E. Fixed Constructions;

F. Mechanical Engineering; Lighting; Heating; Weapons; Blasting;

G. Physics; and

H. Electricity.

The IPC currently contains about 70 000 individual sub-groups, about half the number of the USPC. Section C covers pure chemistry, including inorganic compounds, organic compounds, macromolecular compounds, and their methods of preparation; applied chemistry; and metallurgy. Section C is further divided into 21 classes (See Figure 3.8).

C01	Inorganic chemistry
C02	Treatment of water, wastewater, sewage, or sludge
C03	Glass, mineral or slag wool
C04	Cements; concrete; artificial stone; ceramics; refractories
C05	Fertilisers, Manufacture thereof
C06	Explosives; matches
C07	Organic chemistry
C08	Organic macromolecular compounds; their preparation or chemical working-up; compositions based thereon
C09	Dyes; paints; polishes; natural resins; adhesives; miscellaneous compositions; miscellaneous applications of materials
C10	Petroleum; gas or coke industries; technical gases containing carbon monoxide; fuels; lubricants; peat
C11	Animal and vegetable oils, fats, fatty substances and waxes; fatty acids therefrom; detergents; candles
C12	Biochemistry; beer; spirits; wine; vinegar; microbiology; enzymology; mutation or genetic engineering
C13	Sugar or starch industry
C14	Skins; hides; pelts; leather
C21	Metallurgy of iron
C22	Metallurgy; ferrous or non-ferrous alloys; treatment of alloys or non-ferrous metals
C23	Coating metallic material; coating metallic material with metallic material; chemical surface treatment; diffusion treatment of metallic material; coating by vacuum evaporation, by sputtering, by ion implantation or by chemical vapour deposition, in general
C25	Electrolytic or electrophoretic processes; apparatus thereof
C30	Crystal growth
C40	Combinatorial chemistry

Figure 3.8 IPC Section C, Class Hierarchy.

3.8.4.3 Classification of Markush Structures. A Markush structure is a generic chemical structure defined in a patent claim having at least one common property, function/utility, structural element, or all of the above. Markush structures are named for Eugene Markush, an American chemist who, in 1924, successfully patented a chemical compound containing generic elements. This development was important because it greatly increased the scope of protection obtainable in a single patent. After Markush, chemists could obtain patents on not just single compounds, but generic compounds covering thousands or even tens of thousands of possible substances. Obviously, patents containing Markush structures are difficult to classify under traditional patent classification systems due to the sheer number of codes that could be applied. The USPC and IPC systems have specific rules for the classification of Markush patents. See the respective guides listed in Section 13.

3.9 MAJOR PATENT DATABASES

There are hundreds of public, free, and commercial patent databases and databases containing patent information on the internet. A complete description of even a small subset of them is beyond the scope of this chapter. Below are brief descriptions of the most important databases and search tools. For more information on patent databases, readers are encouraged to consult the resources and websites listed at the end of this chapter.

USPTO: The USPTO website hosts several patent information databases (http://patft.uspto.gov/). The two main databases are PatFT, which contains full-text patents from 1976 forward and images of patent documents from 1790 to the present, and AppFT, which contains full-text published plant and utility applications from 2001 to the present. Chemical structure searching is not supported, and there is no standardized index of chemical names. As of July 1, 2012, PatFT contained approximately 9 million patents and AppFT contained approximately 2.1 million published applications. A TIFF (Tagged Image File Format) plugin is required to view documents.

DEPATISnet: The German Patent and Trademark Office's DEPATISnet patent search system (http://depatisnet.dpma.de) contains patent data from approximately 90 countries and regional patent offices, making it comparable in coverage to the Espacenet system. Search interfaces in German and English are available.

Espacenet: Espacenet (http://worldwide.espacenet.com/) is an international patent database produced by the European Patent Office (EPO). The worldwide collection contains more than 70 million patent documents from approximately 90 countries and regional patent offices, plus more than 2 million non-patent literature references. The EP and WO collections include the searchable full-text of European and PCT published applications. Patent documents in Espacenet are classified under the IPC and ECLA, the European Classification. Patent legal status and family data is sourced from the INPADOC database, which is maintained by the EPO. Year coverage varies by country; countries with the most extensive historical coverage include Germany (1877+), France (1900+), Great Britain (1859+), and the U.S. (1836+). Chemical structure searching is not supported and there is no standardized index of chemical names. Espacenet's machine translation application, created by the EPO and Google, will translate text between English and 13 other languages, including Chinese, French, German, and Spanish. The Espacenet search interface is available in a variety of languages.

Google Patents: Launched in 2006, Google Patents (http://www.google.com/patents) contains full-text U.S. patent documents from 1790 to the present and European patent documents from 1978 forward. Fielded search options include patent number, inventor name, original assignee, date, USPC, and IPC classification. Google Patents is especially useful for retrieving U.S. patent documents from the 19th and early

20th centuries that are not fully indexed in other databases. However, keyword and classification searches may retrieve results that are inconsistent with identical searches in other databases. The reasons for these discrepancies are not known, but are possibly linked to incomplete indexing.

PatentScope: PatentScope (http://www.google.com/patents) is the WIPO's public patent database. When it was launched in 2003, it contained only published PCT applications from 1978 forward. In recent years, WIPO has added numerous national patent collections to the database. As of July 2012, PatentScope contains approximately 2.1 million published PCT applications and 14 million patent documents from more than 30 countries. The most recent additions are the collections of the EPO (1978+), Japan (2004+), and Russia/USSR (1919+). Data from the U.S. and China patent offices is anticipated in 2013. Chemical structure searching is not supported and there is no standardized index of chemical names.

Derwent World Patents Index (DWPI): Produced by Thomson Reuters (Scientific), DWPI is one of the oldest and largest commercial patent databases available on a variety of vendor platforms. It currently contains information on more than 42.5 million patent documents from 47 patent-issuing authorities, and is updated twice weekly. Subject coverage includes pharmaceutical patents from 1963 forward, agricultural patents from 1965, polymer and plastic patents from 1966, chemical patents from 1970 and all patentable technology from 1974.[21] Special value-added features included enhanced titles and abstracts and in-depth chemical and polymer indexing (Derwent Patent Classification).[22] Chemical structure drawings are available back to 1992.

3.10 CHEMISTRY DATABASES AND SEARCH ENGINES CONTAINING PATENT INFORMATION

Chemical Abstracts Service (CAS): Chemical Abstracts Service, a division of the American Chemical Society, produces the world's leading chemistry research database, SciFinder, which includes both CAPlus and REGISTRY. Both databases contain information originally published in *Chemical Abstracts*, which was first published in 1907. From its inception *Chemical Abstracts* covered patents from the major industrialized countries of Europe and North America. At the time of this writing, SciFinder/CAPlus covers patents from 63 patent offices, including Brazil, Russia, China, India, and Korea. Additional selected patents from the early 1800s through 1907 are also available. Patents indexed in SciFinder are selected from more than 35 000 IPC codes and 99 USPC classes. SciFinder offers an array of search options including patent numbers, assignee names, and inventor names, as well as both structure and physical property searching in REGISTRY, reaction searching through CASREACT, and rough Markush searching through the MARPAT database.

Reaxys: Reaxys is a web-based chemistry database launched by Elsevier in 2009 as the successor to the CrossFire Commander client-based search system. Reaxys contains experimental chemical and physical property data and reaction information published in the chemical literature dating back to 1771. Most of the historical data in Reaxys was originally published in Beilstein's *Handbook of Organic Chemistry* (*Handbuch der Organischen Chemie,* 1771–1980 and Gmelin's *Handbook of Inorganic and Organometallic Chemistry (Handbuch der Anorganischen Chemie*), 1817–1975. Organic chemistry information from 1960 forward is obtained from selected journals and patents. Inorganic and organometallic chemistry information is taken from journals only. Approximately 17% of the 4.6 million references in Reaxys are patent documents. Historical patent coverage (mostly organic chemistry) for major patent-issuing countries such as France, Germany, Great Britain and the U.S. ranges from the mid-1800s through about 1981. Current patent coverage in the Patent Chemistry Database component of Reaxys is limited to English-language EP (1978 +), US (1976 +) and WO (1978 +) patent documents in IPC classes C07 (Organic Chemistry), A61K (Medicinal, Dental, Cosmetic Preparations) and C09B (Dyes). Patents for polymers are not indexed unless they are assigned to one of the IPC classes mentioned above. Biosequences disclosed in patents are indexed by name from 2003 onward. Searchable patent indexes include patent number, application number, data of filing, priority date and number, inventor name, patent family data, main and secondary IPC. Prophetic compound and Markush structures are also searchable.

ChemSpider: ChemSpider (http://www.chemspider.com) is an award-winning free chemical information search engine owned by the Royal Society of Chemistry. As of July 2012 it provides access to over 26 million chemical structures, properties and other information from more than 400 data sources. Search options include systematic name, synonym, trade name, registry number, SMILES, InChI, CSID, structure, and properties. Patent data is sourced from Google Patents and SureChem, a proprietary chemistry patent database. Patent coverage includes USPTO, EPO, WIPO/PCT patent documents, and Japanese patent abstracts from the late 1970s forward.

PubChem: PubChem (http://pubchem.ncbi.nlm.nih.gov/) is a free chemical depository and search system operated by the U.S. National Library of Medicine which contains information on the biological activities of small molecules. PubChem consists of three linked databases, PubChem Substance, PubChem Compound, and PubChem BioAssay. At the time of this writing, PubChem Substance contains approximately 100 million records submitted by depositors; PubChem Compound contains 46 million unique structures, including 14 million sourced from EP, JP, US, and WO patent documents, and validated chemical information. Individual PubChem compound summaries include references to related patent documents that are supplied by the depositor of a substance. The Advanced Compound Search allows searching by patent number, which will retrieve all the compounds linked to a specific patent document. Patent documents can be retrieved from a patent office website or SureChemOpen.

SureChem: SureChem (http://surechem.com/) is a suite of patent chemistry databases produced by Digital Science, a business unit of Macmillan Publishers. SureChem data is sourced from the Claims® Global Patent Database, which is produced by IFI Claims®, a subsidiary of Fairview Research. At the time of this writing, SureChem patent coverage includes EP (1978+), JP (1972+), US (1920+), WO (1978+), plus 90 other countries from about 1920 forward. The free version, SureChemOpen, offers structure and keyword searches and links to the Royal Society of Chemistry journals and ChemSpider, but limits the number of searches and does not allow data export.

3.11 RESOURCES FOR DRUGS AND PHARMACEUTICAL PATENTS

Merck Index: The venerable Merck Index is an encyclopedia of chemicals, drugs, and biological substances first published in 1889. The 14th edition, which was published in 2006, contains monographs for more than 10 000 substances and closely related compounds. Substances are indexed by name, formula, therapeutic category, and biological activity and CAS registry number. The online edition, which is updated approximately twice per year, is available on Dialog (File 304), STN (MRCK), Knovel, and CambridgeSoft. Patent information is provided for many compounds. Patent citations include the two-letter country code, number, year of publication, and assignee if known.

FDA Orange Book: The U.S. Food and Drug Administration's Approved Drug Products with Therapeutic Equivalence Evaluations (http://www.accessdata.fda.gov/scripts/cder/ob/default.cfm), commonly known as the Orange Book, contains public information on thousands of prescription drug products, over-the-counter (OTC) drug products and discontinued drug products. The 32nd annual edition was published in 2012. The Electronic Orange Book (EOB) is available as a PDF and searchable database on the FDA website. Searchable data fields include active ingredient, proprietary name, U.S. patent number, applicant name, and FDA application number. Drug product entries include patent number, patent expiration date, and patent use code. The EOB is updated daily for generic drug approvals and monthly for new drug application approvals. Patent information is updated daily also.

Canadian Patent Register: The Canadian Patent Register (http://www.patentregister.ca/), which is maintained by Health Canada, provides access to public information about human and veterinary drugs and their associated Canadian patents. As of December 2011, the Register contained 502 medicinal ingredients and 896 patents. The database is updated nightly and includes data from 1993 forward. Searchable data fields include active ingredient, brand name, Drug Information Number (DIN), and patent number. Individual entries include patent number, patent expiration date and applicant status (owner, licensee).

3.12   PATENT INFORMATION ASSOCIATIONS AND RESOURCES

American Chemical Society (ACS) Committee on Patents and Related Matters (CPRM): One of the main goals of the CPRM is to educate ACS members about the importance of intellectual property issues in chemistry. The CPRM website contains a wealth of useful information about patents including the guide, *What Every Chemist Should Know About Patents.* See http://portal.acs.org/portal/PublicWebSite/about/governance/committees/patents/index.htm.

Patent Information Users Group (PIUG): Founded in the U.S. in 1988, PIUG is an international organization of information professionals interested in all aspects of patent information. It currently has more than 700 members representing 27 countries. A sub-chapter was recently established in China. The majority of members are professional patent searchers working for corporations. However, the membership also includes patent attorneys, patent agents, patent information vendors, licensing professionals, patent information researchers, and librarians. PIUG organizes various meetings, workshops, and an annual conference. The PIUG website, wiki, and newsletter are rich sources of information on patents and patent search tools. See http://www.piug.org.

Confederacy of European Patent Information Groups: The CEPUIG is an umbrella organization for European patent information groups. Members include the UK Patent and Trademark Group, a special interest group of the Chartered Institute of Library and Information Professionals (CILIP), and groups in Belgium, The Netherlands, Germany, Denmark, Sweden, Italy, France, and Switzerland. See http://www.cepiug.org/.

Patent and Trademark Resource Center Association (PTRCA): The PTRCA is a professional association of librarians and library staff who work in the 81 academic, public, and special libraries that are members of the USPTO's network of Patent and Trademark Resource Centers. The PTRC website contains useful guides, teaching materials, and information on historical patents. See http://www.ptdla.org/.

Intellogist: Launched in 2009, Intellogist is a free website for patent information professionals sponsored by Landon IP, a patent research firm located in Alexandria, Virginia. Intellogist features in-depth reports and comparisons of free and commercial patent search tools, an interactive patent coverage map, a library of best practices, and community reports. The Intellogist Blog features articles on a variety of topics written by professional patent searchers at Landon IP. See http://www.intellogist.com.

British Library Patents Collection: The British Library in London maintains one of the largest collections of patent information, much of it historical, in the world. See http://www.bl.uk/reshelp/findhelpsubject/busmanlaw/ip/ippatents/patents.html. Stephen van Dulken, a patent information expert in the British Library Research Service, publishes the Patent Search Blog. See http://britishlibrary.typepad.co.uk/patentsblog/.

World Patent Information: WPI is a peer-reviewed journal launched in 1979 by WIPO and the Commission of European Communities and now published by Elsevier. WPI publishes original articles on all aspects of patent information and documentation, news from the patent offices, reports from patent information conferences and meetings and book reviews.

European Patent Office (EPO): The EPO website contains a wealth of information on patents. In additional to Espacenet and the

European Patent Register, there is the virtual helpdesk for Asian patent information, a quarterly *Patent Information News* newsletter, and special articles on patenting issues. The EPO also hosts an annual Patent Information Conference. See http://www.epo.org.

World Intellectual Property Office (WIPO): The WIPO website offers many useful resources for inventors, patent searchers, and patent attorneys. Some of the most useful materials include the *Handbook on Intellectual Property Information and Documentation*, which includes a complete set of WIPO standards, examples of national numbering systems and documents and patent classification information. The WIPO E-bookshop includes many free publications on intellectual property rights written for independent inventors and entrepreneurs. See http://www.wipo.int.

FURTHER READING

S. R. Adams, *Information Sources in Patents*, De Gruyter Saur, Berlin, 3rd edition completely rev., 2012.

D. Hunt, L. Nguyen and M. Rodgers, *Patent Searching: Tools & Techniques*, John Wiley, Hoboken, N.J., 2007.

M. Gewehr, I. Schellner and K. Hinkelmann, *Japanese-English Chemical Dictionary: Including a Guide to Japanese Patents and Scientific Literature*, Wiley-VCH, Weinheim, 2008.

L.-N. McLeland, *What Every Chemist Should Know About Patents*, American Chemical Society, Washington, D.C., 3rd ed., 2002. (http://portal.acs.org/portal/PublicWebSite/about/governance/committees/WPCP_006903)

C. P. Miller and M. J. Evans, *The Chemist's Companion Guide to Patent Law*, John Wiley, Hoboken, N.J, 2010.

U. S. Patent and Trademark Office, *Handbook of Classification*, U.S. Patent and Trademark Office, Alexandria, Virginia, 2005.

F. J. Waller, *Writing Chemistry Patents and Intellectual Property: A Practical Guide*, John Wiley, Hoboken, N.J., 2011.

World Intellectual Property Organization, *International Patent Classification Guide, version 2011*, World Intellectual Property Organization, Geneva, 2011.

World Intellectual Property Organization, *Inventing the Future: An Introduction to Patents for Small and Medium-sized Enterprises*, World Intellectual Property Organization, Geneva, 2006. http://www.wipo.int/freepublications/en/sme/917/wipo_pub_917.pdf, Accessed on August 30, 2012.

REFERENCES

1. L. H. Baekeland, *Chem. Metall. Eng.*, 1913, **11**, 31.
2. Anon, A. MacDonell (Collachie), *Dictionary of Canadian Biography Online*, ed. J. English and R. Bélanger, University of Toronto and Université Laval, vol. V, 2011, http://www.biographi.ca/009004-119.01-e.php?&id_nbr=2516, Accessed on July 30, 2012.
3. C. Slack, *Noble Obsession: Charles Goodyear, Thomas Hancock, and the Race to Unlock the Greatest Industrial Secret of the Nineteenth Century*, Hyperion, New York, 2002, 51.
4. S. Garfield, Simon, *Mauve: How One Man Invented a Color That Changed the World*, W.W. Norton & Co, New York, 2001, 43.
5. J. Fleischer, *Chem. Eng. News*, 1952, **30**, 239–241.
6. Canadian Intellectual Property Office, *Annual Report 2010–11*, Canadian Intellectual Property Office, Gatineau, Québec, 2011. http://www.cipo.ic.gc.ca/eic/site/cipointernet-internetopic.nsf/eng/h_wr00094.html, Accessed on July 30, 2012.
7. World Intellectual Property Organization, *IP Facts and Figures*, World Intellectual Property Organization. Geneva, 2012, 21.
8. Y. Pope, *CAS Registry^{SM}: The Quality of Comprehensiveness is Not Strained*, EMBL-EBI Industry Program Workshop, Hinxton, Cambridge, UK, December 1, 2010.
9. R. D. Walker, *Patents As Scientific and Technical Literature*, Scarecrow Press, Metuchen, N.J., 1995.
10. K. Makin, "Supreme Court Backs Canadian Firm's Bid to Make Generic Viagra" *The Globe and Mail*, November 8, 2012.
11. European Patent Office, *et al.*, *Four Office Statistics Report*, 2010 edition, Japan Patent Office, Tokyo, 2010, 52. http://www.trilateral.net/statistics/tsr/fosr2010.html. Accessed on August 1, 2012.
12. Burrows, Robert, Pharmaceutical Tablet, US D413972, 1999.
13. M. J. M. A Guillaume and Y. L Lang, Process for Regioselective Mono-tosylation of Diols, US 8,143,432 B2, 2012.
14. U.S. Patent and Trademark Office, Data Visualization Center, Patents Dashboard, July 2012. http://www.uspto.gov/dashboards/patents/main.dashxml. Accessed on July 30, 2012.
15. World Intellectual Property Organization, *Standard ST.9: Recommendation Concerning Bibliographic Data on and Relating to Patents and SPCs*, May 2008, http://www.wipo.int/export/sites/www/standards/en/pdf/03-09-01.pdf. Accessed on August 1, 2012.
16. World Intellectual Property Organization, *Standard ST.3: Recommended Standard on Two-Letter Codes for the Representation of States, Other Entities and Intergovernmental Organizations,*

November 2011, http://www.wipo.int/export/sites/www/standards/en/pdf/03-03-01.pdf. Accessed on August 1, 2012.

17. World Intellectual Property Organization, *Standard ST.16: Recommended Standard Code for the Identification of Different Kinds of Patent Documents*, June 2001, http://www.wipo.int/export/sites/www/standards/en/pdf/03-16-01.pdf. Accessed on August 1, 2012.

18. R. Kaminecki, "A Proximal and a Distal Tip: Not Everyone is Blessed with a Last Name Like Kaminecki". *Dialog Chronolog,* March 2010, http://support.dialog.com/enewsletters/chronolog/201003/. Accessed on July 30, 2012.

19. U.S. Patent and Trademark Office, U.S. Patent Classification System, http://www.uspto.gov/patents/resources/classification/index.jsp. Accessed on August 1, 2012.

20. World Intellectual Property Organization, International Patent Classification, version 2012.01. http://www.wipo.int/classifications/ipc/en/. Accessed on August 1, 2012.

21. Thomson Scientific, *Derwent World Patents Index: Dialog Online User Guide*, Thomson, 2007. http://www.thomsonscientific.com/media/scpdf/dialog_guide.pdf. Accessed on October 30, 2012.

22. Thomson Scientific, *Derwent World Patents Index: CPI Chemical Indexing Guidelines – Indexing of Chemical and Pharmaceutical Patents, 2nd edition*, Thomson Derwent, London, 2001.

III
The Secondary Literature and Specialized Search Techniques

CHAPTER 4

Searching Using Text: Beyond Web Search Engines

ANDREA TWISS-BROOKS

University of Chicago Library, 5730 S. Ellis Avenue, Chicago, IL 60637, US
Email: atbrooks@uchicago.edu

4.1 WHY WEB SEARCH ENGINES AREN'T ENOUGH

There is often a bewildering choice of sources for searching for topics in chemical information using text words. When looking for basic facts or for a specific article reference, Internet search engines (including Google Scholar) that search content on the open Web are often good tools. Freely available Web sites of chemical information, such as spectra, chemical properties, safety information, and more, abound. Using a Web search engine to perform chemically oriented topic searches using text terms can yield some good information, but this is usually not the best method of performing a more comprehensive topic search. When doing a more comprehensive search on a topic, you will need to use specialized tools. Journal publisher Web sites; publisher-produced, subject-specific databases (most of which require that your organization has a paid subscription or that you pay-as-you-go to use); online dictionaries, handbooks and other reference works; institutional or disciplinary repositories of preprints or final author manuscripts;[1] and online collections of e-books provide additional resources for researching a particular chemical topic. Which of these you choose to

Chemical Information for Chemists: A Primer
Edited by Judith N. Currano and Dana L. Roth
© The Royal Society of Chemistry 2014
Published by the Royal Society of Chemistry, www.rsc.org

consult will depend on your particular topic, and a comprehensive search probably requires that you use multiple resources in order to ensure a thorough search.

Topic searches should be a normal part of your strategy for a comprehensive literature review. In-depth topic searches are best performed using one or more subject-oriented databases. The best known and most comprehensive subject database for chemistry is SciFinder, produced by Chemical Abstracts Service,[2] but, depending on your particular research topic, you may need to perform a topic search in additional databases. Other well-known chemistry databases, such as Reaxys,[3] focus more on chemical and physical property or reaction data and may be less useful for text-based topic searching. Other commonly encountered scientific and engineering subject databases that contain references to the chemical literature are Inspec,[4] PubMed,[5] GeoRef,[6] Compendex,[7] and Web of Science.[8] Subjects covered in these databases include physics (Inspec), medicine (PubMed), and earth sciences (GeoRef), as well as related chemical topics such as chemical physics, toxicology and medicinal chemistry, and geochemistry and atmospheric chemistry. If you are interested in business and economic aspects of industrial chemicals, environmental issues related to chemicals, or other tangential areas you may also want to consider selecting a business database and investigating government publications or other types of resources. Coverage of the databases varies not only by subject area but also by types of literature covered, and you should take this into consideration when you are selecting a database to search. If you are having difficulty in selecting or identifying the most appropriate database for researching a particular topic in chemistry or a related subject, consider consulting a chemistry or science librarian or information professional for assistance. If no chemistry or science librarian is available, you can still take advantage of the expertise of librarians by consulting one or more of the numerous library guides for chemistry that may be found on the Web.[9]

One of the most important factors in doing effective topic searching is the critical analysis of your search results. When you do a search, you want to retrieve results that have *relevance* to your research question. *Relevance* is defined in this context as "the ability (as of an information retrieval system) to retrieve material that satisfies the needs of the user".[10] Each researcher typically wants to find all the relevant results for his or her topic. However, when you want to retrieve the maximum number of relevant results, you may need to increase the *recall* of your search, which inevitably leads to retrieval of less relevant results (sometimes known as "false drops") and, therefore, a reduction in the *precision* of the search.[11] In cases where a very comprehensive search is required, you will need to individually review each retrieved reference to identify and eliminate the non-relevant results. Before starting your topic search, you need to have a general sense of the scope and magnitude of your expected search results. Is your topic part of a large research area where there has been extensive research publication? Is your topic in a new or emerging area of study where you do not expect very many publications? The judgment of how comprehensive a search needs to be also depends in large part on the consequences of missing a relevant citation; the cost of repeating a single experiment in the lab because of missing an article may be much lower than that of missing a critical reference for a patent application. One of the most difficult times to know when you have done a good search is when you have few or no results. It may mean that you have identified a fruitful area for new research, or it may just mean that you have not done a good topic search.

Using all the tools at your disposal includes examining the information available as a result of your initial searching efforts. If you have full text of a particularly pertinent journal article reference available, you can always go to the original research report. However, if you are not affiliated with a library that holds an institutional subscription and do not have your own personal subscription to a specific journal title, you may still have access to a considerable amount of information about the article or report. While the initial results of your topic search may be displayed using a short view that often includes the bibliographic information (author, article title, journal title, volume, date, pages, *etc.*), most databases will have at least one viewing option that provides additional information about the publication. These may include an author-provided or other abstract, additional subject terms or keywords that describe the content of the article, and a list of references used by the author in the footnotes or list of works cited. Even if the database you are searching does not include this additional information, many journal publishers provide access to abstracts and references on their publishing Web sites and do not require a subscription or other payment to access this additional information. They generally also offer the option to purchase access to the article, but it is important to check with your library or information center to see what document delivery options exist within your organization before paying to view a paper.

4.2 PRACTICAL APPROACHES TO SEARCHING A TOPIC USING SUBJECT DATABASES

When doing topic searching, it is important to understand the underlying concepts behind the construction of each database system or search engine. How your results are retrieved and presented may depend on a behind-the-scenes ranking algorithm, an underlying subject thesaurus, term mapping functionality that suggests synonyms to use, and the handling of term variants (different spellings, endings, and other variations). Each of these concepts will be discussed in more detail later in this chapter.

In addition to understanding how databases and other tools are constructed and what tools may be available to help construct a topic search in each database, it is also important to understand that chemical topic searching differs from other types of topic searching. Terminology is highly technical and often includes non-alphanumeric characters such as symbols, sub- and super-scripts, Greek letters, and punctuation, which may need to be entered in specific ways when constructing a topic search. Even the most modern search interfaces still rely on indexing done decades earlier, at a time when representation of Greek letters, mathematical symbols, and other special characters were not well handled by computer search systems. Since each database interface treats these special characters differently, you should consult the online help feature of the individual database for assistance, or seek advice from your librarian or a more experienced user.

4.2.1 Controlled Text Terms

Topic terms appear in the title, author- or publisher-written abstract, full text of the article, author-supplied keywords, and additional subject or indexing terms. Most subject databases provide these additional indexing terms that are applied by automated algorithms, by human indexers, or both. These additional indexing terms are usually selected from an official list of terms maintained by the database or index producer and are sometimes referred to as *controlled vocabulary*. In order to perform a comprehensive subject search or to improve the precision of your search, consider using the database's controlled vocabulary, but recognize that use of controlled vocabulary terms does have some limitations. New or emerging areas of research will not be well covered by controlled vocabulary terms since it takes some time for new terms to be adopted into a controlled vocabulary. Records for very recently published journal articles may be included in the databases as provisional or preliminary records before controlled indexing terms have been applied.

When controlled vocabulary is available, the records in the subject database usually display the terms in separate fields. Each individual field or element of the record (accession number, dates of various kinds, journal title, page numbers, abstract, *etc.*) usually starts on a separate

```
PMID- 22165838
OWN - NLM
STAT- MEDLINE
DA  - 20120117
DCOM- 20120309
IS  - 1556-9519 (Electronic)
IS  - 1556-3650 (Linking)
VI  - 50
IP  - 1
DP  - 2012 Jan
TI  - Case series of selenium toxicity from a nutritional supplement.
PG  - 57-64
AB  - INTRODUCTION: Selenium is an essential trace element, but can be toxic in
excess. In May 2008, US FDA reported 201 individuals with adverse reactions to
liquid nutritional supplements containing excess selenium and chromium
...
AD  - The Royal Clinics, Saudi Arabia. barrak_h@yahoo.com
FAU - Aldosary, Barrak M
AU  - Aldosary BM
...
MH  - Adolescent
MH  - Adult
MH  - Alopecia/chemically induced
MH  - Dietary Supplements/*poisoning
MH  - Exanthema/chemically induced
MH  - Female
MH  - Hair/drug effects
MH  - Humans
MH  - Male
MH  - Middle Aged
MH  - Nail Diseases/chemically induced
MH  - Selenium/blood/*poisoning/urine
EDAT- 2011/12/15 06:00
MHDA- 2012/03/10 06:00
CRDT- 2011/12/15 06:00
PHST- 2011/12/14 [aheadofprint]
AID - 10.3109/15563650.2011.641560 [doi]
PST - ppublish
SO  - Clin Toxicol (Phila). 2012 Jan;50(1):57-64. Epub 2011 Dec 14.
```

Figure 4.1   Example of a database record from PubMed containing controlled vocabulary terms (Note: portions of the record were omitted for clarity and are indicated by ellipses).

line and is preceded by a tag or code representing the name of the element or field. An example of a fielded record[12] from the PubMed database appears in Figure 4.1.

Controlled vocabulary terms in Figure 4.1. are preceded by "MH", which stands for Medical Subject Heading, or MeSH. The MeSH vocabulary is a highly developed hierarchical system of controlled terms created by the National Library of Medicine for use in indexing the medical literature. In SciFinder, the additional controlled vocabulary terms appear in the Indexing section of the full record view, as "Concepts" (see Figure 4.2). In this example, the topic of interest that was searched was "natural products for insecticidal applications" and we can see the presence of controlled vocabulary terms like "termiticides" and "insecticides" as well as the concepts "Medicinal Plants" and

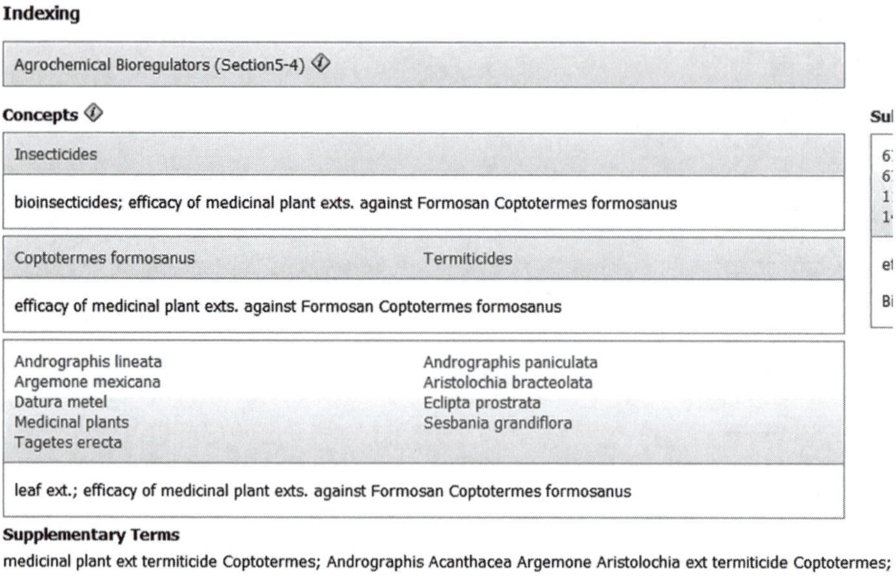

Indexing

Agrochemical Bioregulators (Section5-4) ◈

Concepts ◈   Sul

| Insecticides | | 6: |
| | | 6: |

| bioinsecticides; efficacy of medicinal plant exts. against Formosan Coptotermes formosanus | | 1: |
| | | 1· |

| Coptotermes formosanus | Termiticides | e: |

| efficacy of medicinal plant exts. against Formosan Coptotermes formosanus | | B: |

Andrographis lineata	Andrographis paniculata
Argemone mexicana	Aristolochia bracteolata
Datura metel	Eclipta prostrata
Medicinal plants	Sesbania grandiflora
Tagetes erecta	

leaf ext.; efficacy of medicinal plant exts. against Formosan Coptotermes formosanus

Supplementary Terms

medicinal plant ext termiticide Coptotermes; Andrographis Acanthacea Argemone Aristolochia ext termiticide Coptotermes;

Figure 4.2   Indexing terms from SciFinder record Accession Number 2012:147128
Published with permission of the American Chemical Society.

"efficacy of medicinal plant exts.," as well as a variety of plant species names. It is worth noting, however, that when a more specific term in a hierarchical system is applied to the indexing for an article, the higher level or more general term is not necessarily applied as well. For example, if an article's indexing includes the MeSH term "paclitaxel", the same article would probably not be indexed using the more general term "taxoids". Assuming that you are interested not only in paclitaxel, but, more generally, in the entire taxoid class of compounds, searching for the controlled term "taxoids" would not retrieve the original paclitaxel article. Therefore, it is critical to know the limitations of the controlled vocabulary system used by your database of choice.

Some databases provide assistance for selecting appropriate controlled indexing terms and/or other keywords in the form of browse indexes. Browse indexes allow you to enter a full or partial search term and see what terms are allowed and/or what terms appear in that field (sometimes showing the number of times each term is used in the database). Use of these indexes is extremely helpful, especially when constructing searches using truncation and wildcards (concepts which will be covered a bit later on). Since there is no way to browse all indexing terms used in SciFinder, you can look for appropriate indexing terms to use in an iterative manner. Construct your search using the terminology with which you are already familiar, and then review your preliminary results. Select references that best fit your search topic, and examine the indexing terms used, as well as the words used in the titles and abstracts. Then repeat your search incorporating the appropriate indexing terms to find additional references on your topic.

4.2.2 Using Boolean Logic and Term Adjacency in Text Searching

One advantage of searching in databases that use fielded search terms is that the terms can be combined in precise and predictable ways. Use Boolean operators to limit your topic search in a way that specifies a particular relationship among the search terms used. The most common Boolean operators are AND, OR, and NOT. In a search that uses AND, all the words *must* appear together in each result. In a search that uses OR, each result must contain at least one of the terms entered. The OR operator is often used to search simultaneously for one of several synonyms: for example, "drugs OR pharmaceuticals". The NOT operator excludes a word from appearing in a search result. The AND and NOT operators narrow your search (fewer results are retrieved). The OR operator broadens the search (more results are retrieved). The use of Boolean operators in databases and search platforms is ubiquitous, and more information about applying such operators is available in the help section of each specific database. Some databases supply a structured search form that provides search term boxes connected by Boolean operators in pull down menus. In others, you may need to type the Boolean operators into a search box yourself. In databases that supply a structured search form, do be aware that the order of Boolean operators can affect your results, especially when combining ANDs, ORs, and NOTs in various combinations. One warning: in databases where natural language processing algorithms are used, (*e.g.*, SciFinder), Boolean operators may be ignored or processed in a non-standard way, resulting in unexpected retrieval of search results. In SciFinder,[13] the Boolean OR is replaced by the use of parentheses to indicate search term synonyms. If you wanted to search for "antibiotics used in treating cats" and include the term "felines" as a synonym, you would enter *antibiotics used in treating cats (felines)* and not *antibiotics used in treating (cats OR felines)*. The SciFinder help feature cautions against using Boolean logic at all, and instead suggests using conjunctions and prepositions to convey relationships among terms.

Phrase searching is another common feature of many search interfaces. Phrase searching is particularly useful when a topic you are interested in searching is characterized by a commonly used phrase, *e.g., olefin metathesis* or *Woodward-Hoffmann addition*. Typically, you indicate that you want to search two or more words together as a phrase by enclosing them in quotation marks: for example, "olefin metathesis". Searching by phrase alone can limit your retrieval of relevant results and should generally not be used as the single means of searching for a particular topic. When using a phrase searching approach, it is advisable to examine several answers in your results set for additional terms or variations on the phrase which you may want to search to expand your retrieval of relevant information.

Proximity searching is a technique that provides results that usually contain fewer off-target references than a simple Boolean operator driven search strategy does. The premise is that the closer together words appear in the title or abstract, the more closely related the words are as concepts. Not all databases provide explicit proximity search capabilities, but there are a few common tools that do. EBSCO's databases, including the broad interdisciplinary search tool Academic Search Premier, allow proximity searching. EBSCO uses the letter *w* combined with a number to indicate that references contain the two words or phrases separated by a maximum, specified number of words. For example, if you are interested in finding articles describing the synthesis of ketones, you could enter *synthes* w5 ketone** as a search strategy and retrieve articles with titles such as "**Synthesis** of Triphenylamine-Modified Arylates and **Ketones** via Suzuki Coupling Reactions."[14] In this example, both truncation (discussed further in the next section) and proximity searching were used in conjunction to find references in which either singular or plural forms of the two keywords occur within 5 words of one another. Web of Science offers a similar type of searching with the operator *NEAR/x*, retrieving references in which the two terms appear in either order and within *x* words of one another. Finding the right number of intervening words is somewhat of an art, but a rule of thumb is that a proximity of 5–10 words usually retrieves fairly relevant results.

4.2.3   Word or Term Variants

Unless you are using a database interface like SciFinder with a sophisticated set of behind-the-scenes search algorithms that take into account variations in spelling, word endings, common acronyms/abbreviations, and other word variants, you will probably need to consider how to construct a topic search using some type of lemmatization technique.[15] Different databases provide tools to accomplish this in different ways. Truncation (or stemming) is a lemmatization technique in which a wildcard character is appended to the end of a word fragment to indicate any number of characters following the fragment. For example, in many database interfaces, the truncation character is an asterisk (*), and if you enter *prop** as a search term you will retrieve records containing the words *propane* and *propylene*, among others. You do have to be careful to consider the unintended retrievals that a truncated search term will return; not only will you retrieve the desired chemical terms, you would also retrieve records with the terms *properly* and *propeller*. American and English spelling variants are another problem which can be solved by applying one or more wildcards. In Web of Science, the dollar sign ($) may be used within a word to stand for exactly zero or one additional character. For instance, if you were to enter *colo$r* you would retrieve records with *color* and *colour* present somewhere within the references, but not records with words like *coloner* (which contains two letters between the second *o* and *r*). Other lemmatization characters may be used to stand for any single character; one example is the question mark (*?*) in Web of Science, which can be used to retrieve either *woman* or *women* by entering *wom?n*. The latest version of Web of Science also uses an algorithm to perform some automatic lemmatization, particularly in Topic or Title search terms.[16] SciFinder's algorithms do automated lemmatization as part of that natural language search interface, and the Inspec (physics) and Compendex (engineering) databases on the Engineering Village platform perform "autostemming" which is another type of automatic lemmatization.

Left hand truncation is available in Web of Science and a few other databases, allowing you to construct a topic search in which a variety of prefixes is possible. Suppose you wanted to find articles in Web of Science on the toxicity of uranium. If you enter **toxicity AND uranium* (the asterisk * is the wildcard for left truncation) in Web of Science instead of *toxicity and uranium,* you will retrieve many additional references. This is because you will retrieve references not only containing the term *toxicity*, but also *radiotoxicity, chemotoxicity, genotoxicity, cyctotoxicity, nephrotoxicity, etc.* Left hand truncation is also an approach to doing preliminary searching on classes of substances as a topic (although a true substance search usually requires substructure search techniques that will be covered in another chapter in this book). For example, searching **imidazole* will retrieve any reference with substance ending in -*imidazole* (*e.g., benzimidazole* or *N-vinylimidazole*). In Web of Science (and in some other databases) you may combine left and right truncation in the same search term, *e.g.,* **isotop**. This search would retrieve references containing such terms as *radioisotopes, monoisotopic,* and *nonisotopically.*

4.3   SEARCHING FOR AUTHOR NAMES

Searching for a specific author or authors is a common technique when you know that a particular researcher or research group has published in your area of interest. In order to find all the publications by a particular author, there are some simple but effective rules that you should keep in mind. Database records for references may or may not include the author's complete first name and any middle initials, so check the database help to see if you should use only initials for searching. Some authors' last names, or family names, can be tricky to search, in particular compound or hyphenated last names or names that use diacritics or other special characters. In databases that offer a browse index option for the author name field you can check for variants. In Figure 4.3, the searcher is looking for Paul von Rague Schleyer in the Reaxys database and has typed in part of the author's last name and clicked the browse index button (indicated by a circle). You can see that Paul von Rague

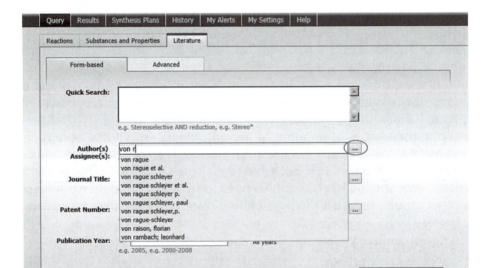

Figure 4.3   Expansion of a browse index for the Author(s)/Assignee(s): field in a Reaxys record – www.reaxys.com. Published with permission of Elsevier. Copyright © 2013 Reed Elsevier Properties SA. All rights reserved.

Schleyer's name has several variations in spelling that would have be taken into account.

Authors sometimes change their names at different points in their careers, for example, after a change in marital status. Researchers with common last names, *e.g.*, Jones or Lee, will also present a challenge in searching. In the case of very common last names, there are often two or more researchers active in an area of research and having the same last name and even the same first/middle initials. In these cases an author search combined (using Boolean AND or NOT) with a topical term or an institutional affiliation can serve to distinguish among various researchers with similar names.

Many articles are written by multiple authors. Databases do not always include or index ***all*** the co-authors listed on a paper, particularly if there are an unusually large number. Author indexing practices in different databases have also changed over time, so it's especially important to understand which author information is included in a database if you need to do a historically comprehensive search.

Recently, there have been efforts to address the ambiguity in author names by promoting the development of an author or researcher identification system. One of these efforts is the Open Researcher & Contributor ID (ORCID) whose aim is "creating a registry of persistent unique identifiers for individual researchers and an open and transparent linking mechanism between ORCID, other ID schemes, and research objects such as publications, grants, and patents".[17] Many major scientific society and commercial publishers are sponsors of ORCID and it is expected that they will incorporate ORCID into their publication process, linking research articles to specific individuals. If ORCID and other similar registration systems become widely established and inter-linked, many of the challenges for searching individual author names could be addressed. Until such time as a single researcher registration system is widely adopted, you will have to consider not only variants in author names, but also the possibility of multiple researcher identifiers.

4.4 SEARCHING FOR REFERENCES CITED IN OTHER PUBLICATIONS

You can leverage the searching that other researchers have done in a specific topic area by extending your searching techniques to include citation searching. Citation searching offers an indirect method for searching a particular topical region of the literature. Since authors typically cite among their references in articles the most pertinent literature, you can use a core article as a starting point. This core article is termed the *cited reference,* and all the other articles that include the core article in their list of references are the *citing references.* By searching for later published articles (*citing references*) that include a reference to the core article you select (*cited reference*), you can follow the connections in the networks of articles through their cited references. In some cases, this may be the single most effective way of searching, particularly for topics which defy easy characterization by keywords.

Several databases (including SciFinder and Scopus) now offer citation linking as a standard feature, but the deepest chronological coverage of cited references is still that which is produced by Thomson Reuters, Web of Science.

4.4.1 Performing a Citation Search

While many databases now offer cited/citing reference linking, how you find citing references is different in each interface. In SciFinder, you must first search and find the core article reference and then follow the *Get Citing* link in the full record view for that article. This will provide a list of all the articles indexed in SciFinder that include your core article in their list of references. In Scopus, you can find citing references directly from the search results display in a column labeled *Cited by.* Each article in the list of results will display a number indicating how many citing references there are for that particular article. Clicking on the number in the *Cited by* column will produce a list of these citing references. The construction of a citation search within Web of Science is done a little differently. The full Web of Science includes over one hundred years of citation information and the consistency with which a particular reference of interest has been cited in the literature may vary quite a lot over time. Web of Science addresses this variation by providing a separate Web search form labeled *Cited Reference Search.* With the exception of the most prolific authors, use the author's name and year for the selected core article as a starting point. (If you have a very prolific author to search, you will need to also include the abbreviated journal name to reduce the number of matching publications retrieved.) Your search of author and year will generate a list of matching references, including any variations that have been introduced by other authors, including any errors in citing that have occurred. You may also want to try several different cited reference searches, leaving out a different piece of information (journal title, year) each time since each part of a reference might have been incorrectly cited by one or more other authors. An illustrative example of a cited reference search from Web of Science is shown in Figure 4.4.

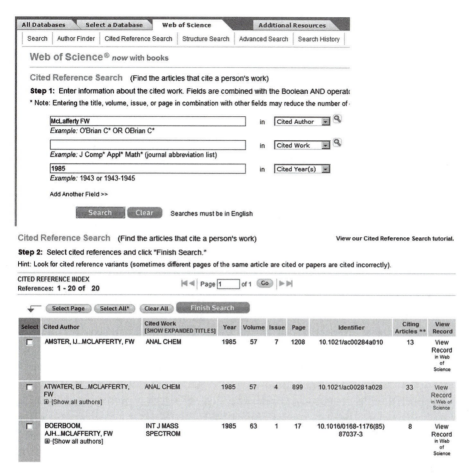

Figure 4.4 Cited reference searching in Web of Science.

Select all the likely matches from the list of matching references to increase your recall of relevant citing articles. You may also need to consider whether an author's name might be represented in different ways in citing articles. As we saw when searching for author names, compound names and hyphenated names are particularly problematic, since spaces and hyphens may have been included or omitted in various combinations by different authors citing the original work or last names may have been misspelled. Since reference styles vary from journal to journal, the same article by the same researcher may or may not have full first names and/or middle initials included in a reference. Once you have selected all the matching citations for your article of interest, complete the search, and you will see a list of references to review.

4.4.2 Determining Citation Impact

Determining the importance of a particular article or researcher is necessarily a subjective exercise, but there are some quantitative measures that you may want to consider when evaluating your search results. If one assumes that a paper that is cited by many other researchers in the field is more important that a paper that is not cited at all, or only cited by a few other authors, the total citation count can be used as a guideline. This count is readily available in several databases, including SciFinder and Web of Science, and you can usually sort the list of results in descending order of the number of citations. But what if there is a brand new paper that has not been published long enough to be included in the list of cited references in others' work? You can make an assumption that an author whose work has been highly cited in the past is influential in the field and by extension her current work is likely to be similarly influential. How do you decide if an author is generally highly cited and therefore influential? While many such citation metrics have been proposed,[18] the most commonly encountered author based metric is the *Hirsch-index* or *h-index*.[19] An author with an *h-index* of *h* has published at least *h* papers, each of which has been cited in other papers at least *h* times. Authors with higher h-indexes may be thought of as more influential than those with lower h-indexes. The h-index, like many other quantitative methods intended to measure quality or importance, has been subject to discussion and criticism.[20,21] Scientists who are early in their publishing careers are at a disadvantage compared to more established authors. The h-index can be manipulated by authors through excessively citing their own previous work (self-citations), although self-citations can be excluded to produce a modified h-index. Variations in citation patterns and publishing activity in different fields are not factored into the calculation of h-index, so authors should not be compared across different fields of research. Some databases that provide h-index information do not include all types of publications; citations to books, conference proceedings, *etc.*, may not be taken into account. The extent to which specific journals are covered also varies from database to database. One database may index every single article in each issue and another database index articles selectively, or the years of coverage of the journal may vary among databases. Therefore, the h-index is not an absolute value, but depends on the database that was used as the source for calculating the value of the h-index.

If an author's h-index is not explicitly provided for you, it is fairly easy to calculate yourself, provided the database includes citation counts. Perform an author search, and sort the results in decreasing order by the number of citations (most databases we have covered in this chapter allow you to sort by "times cited"). The number at which the rank in the results list is approximately equal to the number of citations for that paper provides the value for the h-index. There are also some tools available for you to use in determining an author's metrics, including h-index.[22] Bear in mind, however, that each database only contains the bibliographies of those articles that it indexes; therefore, different tools may report disparate h-indexes for the same researcher.

4.5 CONCLUSION

With the tips, tricks and other information provided here you should be well on your way to exploring text searching beyond Web search engines. The references included here provide a rich source of additional detail on effective text based searching. Effective and efficient text searching is a set of skills that should reduce frustration in searching and save you time in lab and in the library.

REFERENCES

1. Examples of disciplinary repositories include arXiv, an article pre-print in physics http://arxiv.org and PubMedCentral, a repository of final author manuscripts of articles in biomedical sciences http://www.ncbi.nlm.nih.gov/pmc/. Accessed on August 7, 2013.
2. An excellent book on the ins and outs of SciFinder is D. Ridley, *Information Retrieval: SciFinder*. 2nd edition, Wiley, Hoboken, NJ, 2009.
3. http://www.reaxys.com. Accessed on August 7, 2013.
4. http://www.theiet.org/resources/inspec/. Accessed on August 7, 2013.
5. http://www.ncbi.nlm.nih.gov/pubmed/. Accessed on August 7, 2013.
6. http://www.agiweb.org/georef/. Accessed on August 7, 2013.
7. http://www.ei.org/compendex. Accessed on August 7, 2013.
8. http://thomsonreuters.com/web-of-knowledge/. Accessed on August 7, 2013.
9. There are lots of choices for library guides available, but here a few to get you started:
   http://ucsd.libguides.com/chemistry. Accessed on August 7, 2013.
   http://libguides.northwestern.edu/chemistry. Accessed on August 7, 2013.
   http://guides.library.ucsb.edu/chemistry. Accessed on August 7, 2013.

10. http://www.merriam-webster.com/dictionary/relevance. Accessed on August 7, 2013.
11. C. D. Manning, *Introduction to Information Retrieval*, Cambridge University Press, Cambrige, 2008, p. 166. "Precision is the fraction of retrieved documents that are relevant" and "Recall is the fraction of relevant documents that are retrieved."
12. http://www.ncbi.nlm.nih.gov/pubmed/22165838, Record from PubMed database provided by the National Library of Medicine. Accessed on August 7, 2013.
13. A. B. Wagner, *J. Chem. Inform. and Model.*, 2006, **46**, 767–774, doi: 10.1021/ci050481b.
14. J. Wang, C. Yang, H. Peng, Y. Deng and X. Gao, *Synth. Comm.*, 2011, **41**, 832–840. doi:10.1080/00397911003706982.
15. Oxford English Dictionary Second edition, 1989; online version March 2012 http://www.oed.com/view/Entry/107195#eid39659170 Lemmatize, v. is defined as "To sort (words as they occur in a text) so as to group together those that are inflected or variant forms of the same word". Accessed on August 7, 2013.
16. Some examples of the automatic lemmatization of American and British spelling and term variations listed in the online help for Web of Science are: *sulfur/sulphur, ionized/ionised*, and *metalizing/metallising*.
17. http://about.orcid.org/. Accessed on August 7, 2013.
18. http://www.harzing.com/pop.htm. Accessed on August 7, 2013.
19. J. E. Hirsch, *Proc. Nat. Acad. Sci.*, 2005, **102**, 16569–16572. http://arxiv.org/abs/physics/0508025. Accessed on August 7, 2013.
20. http://en.wikipedia.org/wiki/H-index. Accessed on August 7, 2013.
21. R. Costas and M. J. Bordons, *Infometrics*, 2007, **1**, 193–203.
22. http://www.chemconnector.com/2011/04/23/calculating-my-h-index-with-free-available-tools/. Not mentioned in this ChemConnector blog item are a few others, including Quadsearch http://quadsearch.csd.auth.gr/index.php?lan=1&s=2 and Scholarometer http://scholarometer.indiana.edu/. Accessed on August 7, 2013.

CHAPTER 5

Searching by Structure and Substructure

JUDITH N. CURRANO

Chemistry Library, University of Pennsylvania, 231 S. 34th St., 5th Floor, Philadelphia, PA 19104-6323, US
Email: currano@pobox.upenn.edu

5.1 SEARCHING BY STRUCTURE

Many chemists think in terms of structure rather than using textual representations of the substances that they study. Reaction schema are laid out using structures; and ChemDraw figures, as well as crystal structures, efficiently demonstrate the structure of a new molecule or complex. These structures and reactions are indexed by database creators and can be used as entry points to the primary literature. Chemical structures have been represented in a machine-searchable manner for fifty or sixty years, but, prior to the 1980s or 1990s, the complex method in which they were represented in the secondary sources made it difficult for anyone but a trained information professional to search using them.[1] However, in recent years, graphical user interfaces have made structure searching accessible to anyone who is able to draw a structure using ChemDraw or a similar structure editor, and structure search interfaces have popped up everywhere from the Aldrich Catalog[2] to the CAS Registry, *via* SciFinder[®3] or STN[®].[4] This chapter will discuss techniques for performing effective structure searches, as well as methods of

Chemical Information for Chemists: A Primer
Edited by Judith N. Currano and Dana L. Roth
© The Royal Society of Chemistry 2014
Published by the Royal Society of Chemistry, www.rsc.org

searching for substances containing a desirable core or fragment. Readers interested in learning more about historic methods of representing chemical structure for information retrieval can refer to Wendy Warr's 2011 review article, which synthesizes the important changes in methodology and technology from the 1950s to the present day.[1]

The main problem with structure searching today is that many information providers have engaged in parallel development of structure search tools, resulting in a wide variety of interfaces and systems that have some commonalities but have even more differences. Because of the extreme variability of the tools available, we will take a largely theoretical approach to this subject, outlining techniques that are applicable in many systems and presenting search examples that employ these techniques in widely-used resources, such as SciFinder®, the CAS Registry *via* STN®, and Reaxys®.[5] We will also highlight unique methods of searching that may only be available in one or two systems. It is important to consult a resource's help files and acquaint yourself with its structure editor prior to beginning a search, and it is critical to remember to select a tool, first, based on the appropriateness of the content to the query and then based on the search options available. Finally, because interfaces change rapidly, we will avoid the use of screen shots and discuss the use of the various techniques in each tool in more general terms.

5.1.1   Identifying Substances for Information Retrieval

Searching for information about substances requires different access points than performing a text-based search for information on a topic of interest. While it is possible to type the name of a substance, particularly a common substance, into a search engine like Google and retrieve some data and articles about it, it is generally better to begin in a chemistry resource, using one of five specialized identifiers for the substance in question: chemical name, formula, CAS Registry Number, International Chemical Identifier, and structure. The first four of these are text-based methods; the fifth, structure, will be discussed in the next section.

Chemical Name

At first glance, chemical name seems to be the easiest and most straightforward way of searching for information about a substance. One generally knows the chemical's name, and inputting the name requires only basic text searching skills. However, searching by name carries with it a many-to-one problem; there are many different names for a single substance. In addition to the systematic names that a chemical may have, assigned according to rules laid out by the International

Union of Pure and Applied Chemistry (IUPAC) and Chemical Abstracts Service (CAS), respectively, each substance may have numerous common, historic, trade, or brand names. In order to retrieve a substance, it is necessary that the name that you enter into the search system be found in that substance's record. Therefore, you may need to enter several different synonyms before the one is found to match.

Formula

Searching by formula presents a different set of advantages and disadvantages than name searching does. Molecular formula is the most common type of formula to use in information retrieval, although some sources like the CAS Registry *via* STN® will allow one to perform a search using the ratio of two or more types of atom, allowing an approximation of an empirical formula search. While name searching carries a many-to-one problem, searching by formula presents the opposite, one-to-many problem, in that a single formula may represent the composition of several different substances, especially when searching for fairly common organic substances. You must then browse through a variety of hits to locate the substance of interest or use another piece of information, such as the substance's name, in order to identify the correct entity.

CAS Registry Number and International Chemical Identifier (InChI)

It is obvious that a unique identifier would solve the problems of unambiguously identifying each substance, so, in 1965, CAS introduced the CAS Registry Number.[6] This number is very similar to a United States Social Security Number: it is a unique, one-to-one identifier for a substance, it consists of three sets of digits separated by hyphens, and it carries no information whatsoever about the composition or structure of the substance. For example, the CAS Registry Number for water is 7732-18-5. CAS Registry Numbers are assigned when a substance is first published in the literature or patented, although individuals may also call CAS and request registration for a substance,[7] and they are proprietary, meaning that an individual who wishes to use them in a database must request permission or license the numbers of the substances that he or she wishes to include. As a result, not all databases are searchable by CAS Registry Number, although most comprehensive chemical information tools are. It should be noted that, when the Registry Numbers of the same substance in different sources do not match, the *Chemical Abstracts* Registry file should be the final authority.

In around 2000, IUPAC decided to create a second unique identifier, which they dubbed the International Chemical Identifier, or InChI. The goal of this project was to "develop a set of algorithms for the standard representation of chemical structures that will be readily accessible to chemists in all countries at no cost. The standard chemical representation could be used as input into existing and newly developed computer programs to generate a IUPAC name and a unique IUPAC identifier."[8] InChIs are non-proprietary, and individuals may create them using freely-available InChI generators, several of which are linked from the InChI Web site (http://www.iupac.org/inchi). However, they are more complicated to write than CAS Registry Numbers (the InChI for water is 1S/H2O/h1H2); so, one frequently finds them used on the back end of journals and databases, rather than typed into search boxes.

The down side to using either of these forms of identification is that most people do not know them off-hand. A CAS Registry Number may be located in the CAS Registry file or by searching other tools using a different identifier, while one must input a structure into an InChI generator in order to get an InChI.

5.1.2 When to Perform a Structure Search

Chemical structure is the third unique identifier for many molecules. Since it is a one-to-one identifier for a substance and is something that most chemists are comfortable reading and generating, it is one of the best ways of finding a specific substance in the literature. However, as with all methods of identification, structure searching comes with its drawbacks. It can be very cumbersome to draw a large structure, such as a cage compound or a complex natural product. Coordination compounds are represented inconsistently in the literature; since the connections between them are coordinations and not bonds, different resources have different methods of representing them, requiring flexible search parameters in order to retrieve all of the applicable information. Polymers have problems all their own, which will be addressed in a later chapter of this work. Finally, many information resources represent proteins and nucleotides by only their sequences, meaning that it can be difficult to search for these substances by their structure or a portion of their structure. In order to better understand the strengths and weaknesses of structure searching, it is first important to understand how structure searching works.

5.1.3 Behind the Scenes with Structure Searching

Historically, structures in databases have been converted from their graphical form to a table of information describing their atoms and the connections between them. Take, for example, the following substituted furan (Figure 5.1a). Each atom in the substance is assigned a node number (Figure 5.1b).

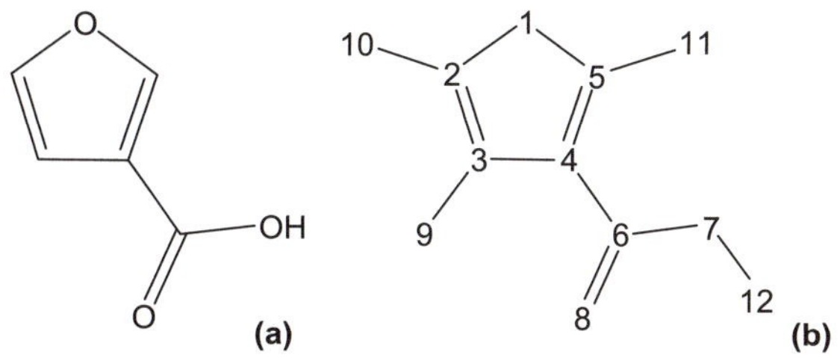

Figure 5.1 (a) A substituted furan. Hydrogen substitution is assumed on the three unsubstituted carbon atoms in the ring. (b) The same substituted furan, with each atom represented by a node number.

Table 5.1 A redundant connection table describing the substituted furan in Figure 5.1. Note that each connection is described twice, in the entries for both atoms that participate.

Node	Atom Represented	Connection Type	Connected Node
1	O	–	2
		–	5
2	C	–	1
		=	3
		–	10
3	C	=	2
		–	4
		–	9
4	C	–	3
		=	5
		–	6
5	C	–	1
		=	4
		–	11
6	C	–	4
		–	7
		=	8
7	O	–	6
		–	12
8	O	=	6
9	H	–	3
10	H	–	2
11	H	–	5
12	H	–	7

The connection table takes each node and lists the identity of the atom represented by that node, the other nodes to which it is connected, and the types of connections used. There are two types of connection tables: redundant, and non-redundant. Table 5.1 illustrates a redundant

Table 5.2   A non-redundant connection table for the substituted furan in
Figure 5.1. Each connection is described exactly once.

Node	Atom Represented	Connection Type	Connected Node
1	O	–	2
2	C	=	3
3	C	–	4
4	C	=	5
5	C	–	1
6	C	–	4
7	O	–	6
8	O	=	6
9	H	–	3
10	H	–	2
11	H	–	5
12	H	–	7

connection table for the substituted furan above, in which every connection that a node makes is listed in its section of the table.

Table 5.2 shows a non-redundant connection table for the same
substance. In it, every connection is displayed exactly once.

The way in which a connection table is built and used varies from
vendor to vendor; again, Wendy Warr's 2011 article does a nice job of
explaining the use of connection tables in a variety of resources.[1] Basically,
when you perform a structure search, the database generates a connection table for the substance drawn into the structure editor. This
connection table is then compared to the connection tables of the substances in the database, and those that match are retrieved as hits. Since
most, if not all, structure editors now have graphical interfaces, it is not
necessary for a chemist to learn the ins and outs of connection table
generation; however, a basic understanding of what the database is
doing can assist with an overall understanding of the search process and
improve search results.

5.2   SEARCHING BY SUBSTRUCTURE

The term "substructure" means exactly what it appears to mean; it is a
segment of a structure. A substructure search for substances, therefore,
is akin to a truncation search for words. When one performs a substructure search, one enters a portion of a molecule into the search
system and retrieves substances that contain that fragment, or "substructure." A visual analogy would be drawing the substructure template on a sheet of glass and superimposing it on every molecule in the
database. Only candidates in which the template exactly overlaps a

portion of the structure will be retrieved as search hits. In actuality, the search system is generating a connection table for the substructure drawn, comparing it to the connection tables of all of the substances in the database, and retrieving those in which the same connections appear. Unlike in an exact structure search, it is not necessary for the connection tables of the query and database substances to be identical; instead, the connection table of the query structure must form a subset of that of the database substance to be retrieved.

5.2.1 Introduction to Substructure Searching

A substructure search can be an open-ended search, in which a searcher hopes to locate a large or focused set of molecules that have a particular framework, or it can serve as a Boolean "OR" search for a few specific substances that all have certain structural elements in common. This chapter will outline general techniques that can be used in either case to retrieve sets of organic or inorganic molecules or coordination compounds. Chapter 9 will show how one can apply substructures to reaction searches. As previously mentioned, there are other methods of searching for proteins and nucleotides; these will be dealt with in Chapter 10.

It seems relatively easy to perform a substructure search: you simply draw a fragment of a molecule and click the "search" button of a database. The challenge comes from developing a substructure with the appropriate degree of specificity; as with any other type of search, one wants the system to retrieve absolutely all of the desired information, while, at the same time, filtering out all irrelevant records. Depending on your end goal, you will want to make the substructure more or less detailed. A search designed to locate four distinct molecules in a single search without retrieving any additional substances is a highly directed search, requiring a very detailed substructure. If, however, you wish to prove that nobody has ever made a molecule of a particular type, you must start with a very general framework and perform an exhaustive search. The first type of search will take a fair amount of time to construct and input but will yield fewer and more relevant results; the second is a quick search to compile and run, but the results will require considerable filtering for relevance. The art of substructure searching comes from balancing the length of time spent constructing the search with the length of time that one wants to spend evaluating the results. In other words, a savvy searcher will always attempt efficiently to build a search that gets everything that he or she wants, with few undesired results.

The other main challenge in substructure searching also closely resembles one of the main problems with searching by text. You will need to guess how a structure of interest has been represented in an article or a database. This is particularly challenging for organometallics and other coordination compounds, where there are no clear conventions in representing bonds and coordinations between atoms. We will deal with this unique challenge in Section 5.3.

5.2.2   Basic Procedure for Substructure Searching

There are four basic steps to constructing an effective substructure.

1. Identify the portion of the molecule(s) that is of chemical interest.
2. Determine whether the stereochemistry, geometry, and bond order of each connection in your structure are fixed or flexible.
3. Decide where and how your substructure may be further substituted.
4. Determine the topology of each atom and connection in the structure.

At each step, it is imperative to remember that you are thinking like a database, rather than like a chemist, which can initially be challenging. You may be starting with a particular substance or transformation in mind, but you need to be careful not to exclude other, potentially interesting or useful results. Therefore, a little extra thought at the outset can mean the difference between one hundred highly relevant hits and ten thousand hits that have little to do with your substance of interest. We will look at all four steps in detail. Note that SciFinder®'s substructure algorithm is more permissive than many others; when you are using this tool, it is helpful to analyze your results by precision to eliminate those that are only loosely related to the substructure that you have drawn.

Identify the portion of the molecule(s) that is of chemical interest

This seems like the easiest part of the entire substructure search process, but its simplicity may be deceptive. It takes little imagination to realize that, if you are trying to find all molecules that incorporate a fragment that you know how to make, your initial structure should include that piece (Figure 5.2). If, however, you are trying to find substances that more distantly resemble your structure of interest, creating the initial substructure is slightly more challenging. You need to be sure to ask yourself exactly how much of the molecule you need to retrieve all of the substances that could possibly be of interest to you with a minimum of undesirable structures, or "false hits". Examine the molecule atom by atom, paying special attention to heteroatoms and metals that may safely be replaced by other elements.

   If you are trying to use a single search to find several molecules with similar cores, you must proceed in a slightly different manner. In this

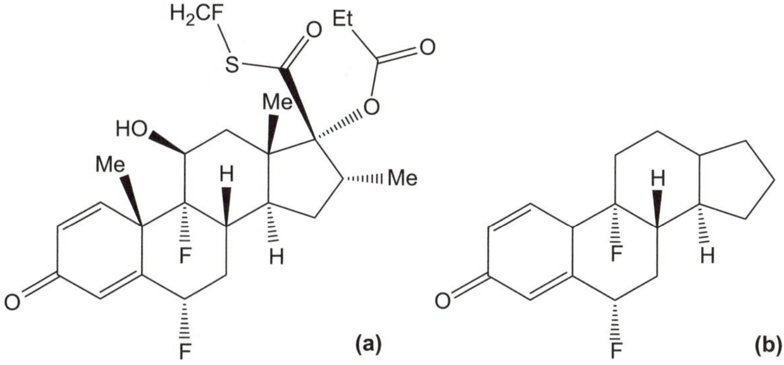

Figure 5.2 (a) The structure of fluticasone propionate. (b) A substructure for that could be used to locate substances similar to fluticasone propionate, focusing on the 6,6,6,5 ring system and four of the attachments.

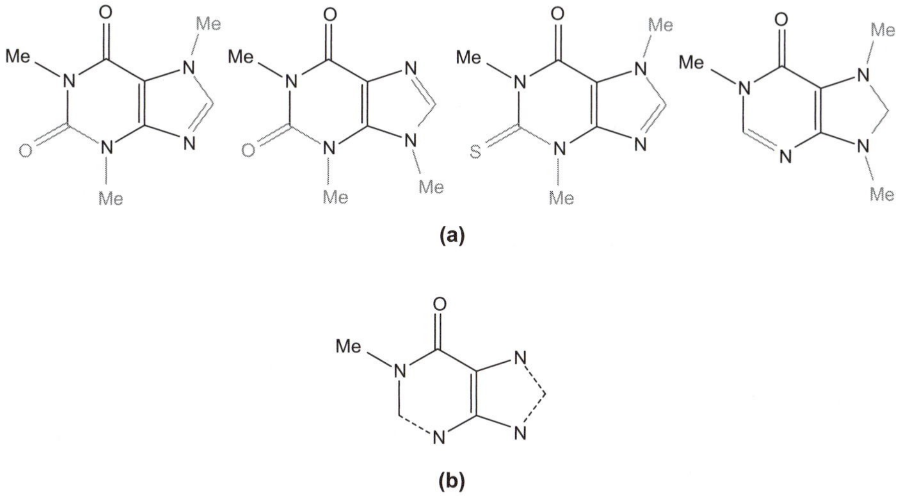

Figure 5.3 (a) The structures of caffeine and three structurally similar substances. The black sections are common to all of the molecules, and the gray parts vary. (b) A substructure that could be used to retrieve all four caffeine-like molecules.

case, you want to examine the molecules and see what segments they all have in common (Figure 5.3a). This will become the basis of the substructure (Figure 5.3b).

Determine whether the stereochemistry, geometry, and bond order of each connection in the structure are fixed or flexible

Once you have determined and drawn the section of interest, you need to make sure that it has the right degree of specificity. The first thing to do is to examine all potential stereocenters in the substance and determine how important it is to have a particular configuration at each. Most search systems will allow you to specify the stereochemistry of a center with wedge and hash bonds, so, if you care greatly about the configuration of the center, use the correct bond types to indicate the directions in which the atoms are pointing. If, however, you are open to any stereochemistry, you will want to use plain single bonds rather than stereo bonds. For some systems, such as SciFinder®, this will suffice; however, for others, like Reaxys®, you will want to make sure that the stereo search is turned off for the best results.

One thing that generally startles scientists new to substructure searching is that, in many systems, the search appears to completely ignore double-bond geometry. Geometry is treated in a manner very similar to stereochemistry by most search systems. If you use a plain double bond, some search algorithms, like SciFinder's®, allow a Z configuration to be replaced by E geometry. These systems have special "double steric" double bonds, which should be used to fix the geometry when it is important.

Finally, after determining the importance of stereochemistry and geometry, you will need to examine every connection in your structure and decide whether the bond order is fixed or flexible. In most cases, you will choose to search for the bond order as it appears in the substance on which you are basing your search. However, at some times, it may not matter whether a bond is single or double. Most search systems have variable bond-order bonds that can be used in such situations. The most common, a dashed bond, is used to represent single, double, or triple bonds, but some systems, like Reaxys®, have bond orders that are single/double or double/triple, if you wish to be more specific. You should not generally use variable bonds to indicate aromatic rings or ring systems; the search algorithms are now intelligent enough to recognize an aromatic system, and they will search for all variations. A few systems, like the Cambridge Structural Database,[9] have special aromatic bonds that you can use, as well. Bond order between a metal and a ligand in organometallics and other coordination compounds is represented differently in every resource, and you should refer to the help files and training information for the resource that you wish to use. Coordination compounds will be discussed further in Section 5.3.

Define the type of substitution permitted at each position

This is frequently the most difficult part of the search, and it is the step that really allows you to tailor the specificity of your search. Here, it is critical to remember that a substructure is a template and that only those structural elements explicitly drawn or added to the substructure will be

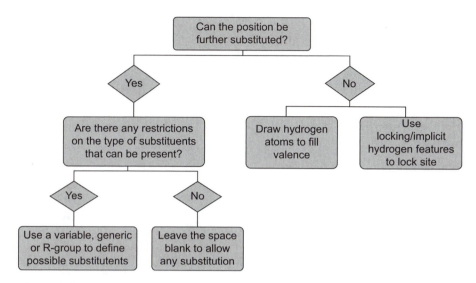

Figure 5.4 A flow diagram representing questions that can be asked about atoms whose valence is not filled.

retrieved in a hit structure. This can be difficult for organic chemists, who are accustomed to assuming the presence of hydrogen in a drawn structure. When performing a substructure search, any atom whose valence is not fully satisfied may have additional substitution of any kind. In some cases, this will be suitable, but in other cases you can get a more focused set of results by making your structure a bit more specific, asking a series of questions about every atom that could possibly form further bonds (Figure 5.4).

Can the position be further substituted?

This is a fairly straightforward question. If a position cannot have additional non-hydrogen substitution, you will need to fill all available positions with hydrogen atoms. Depending on the database that you are searching, you may have two possible ways to do this: draw hydrogen attachments into the structure, or use special features of the structure editor to restrict an atom to hydrogen-only substitution. The first method will work in all search systems, but it has certain drawbacks; some systems have limitations on the number of atoms that they will screen in a structure. The second method requires an in-depth under-standing of the more advanced features of your structure editor. Reaxys® will allow you to specify the total number of hydrogen attachments, and the REGISTRY file on STN® will even allow you to set a minimum and maximum number of hydrogens at a position.

SciFinder® does not have these capabilities, but it will allow you to lock out additional substitution by clicking on an atom with a locking tool. If SciFinder® gives you a "results too general" error, you can frequently narrow the preliminary screen by drawing hydrogen atoms instead of locking out substitution.

Are there any restrictions on the types of substituents that may be present?

If a position can be further substituted, you need to begin asking what kind of substitution that site may have. If you do not really care what is attached, you should leave the position blank, since that places no restrictions on your template and will therefore allow anything. It is when you want to allow non-hydrogen substitution but restrict the types that are permitted that things begin to get interesting. Section 5.2.3.1 deals with methods of limiting the substituents that may attach to the parent structure.

Determine the Topology of Each Atom and Connection in the Structure

Topology refers to whether an atom or bond is in a ring or a chain. Generally, when thinking as a chemist, things that look like rings are rings, and things that look like chains are chains. From the perspective of a database, the first is definitely true. The only possible topology for a cyclic structure is ring topology. However, the way in which the database handles things that look like chains varies greatly from resource to resource. When you use the STN® Web structure editor to draw a chain, that bond is automatically assigned chain-only topology, which means that these bonds can never be part of a ring. However, if you use SciFinder® or Reaxys® and draw something that looks like a chain, the database will allow these connections to have either ring or chain topology, and you may very well find yourself looking at hit structures that have your "chain" atoms as part of an isolated or fused ring or ring system.

It is important to understand the defaults for topology in your search system of choice. Most systems will allow you to specify a chain that you draw to be either chain-only or ring/chain topology, so, it is important to examine the tools available for doing this. In some systems, like SciFinder®, a "lock out" tool allows you to lock out ring formation. In other systems, like STN® or Reaxys®, you get to the topology tools through a right-click or an atom or bond attributes menu. Some systems will allow you to set topology on both atoms and bonds; others allow you only to set the topology of connections between two atoms. It is also important to note that SciFinder will only allow you to lock out ring formation across an entire chain, while other systems allow you to set the topology of individual bonds in the chain.

In addition to setting the topology of substructures that look like chains, it is important to pay attention to rings, as well. Unless otherwise specified, a ring may be part of a larger ring system. Methods of isolating entire ring systems or portions of ring systems, as well as other advanced uses of topology tools are laid out in Section 5.2.3.4.

5.2.3 Advanced Substructure Techniques

5.2.3.1 Generics and R-Groups. Assume that you wish to search for a halogenated benzene ring. You wish to allow any halogen to be present, but you want the halogen to be the only substituent on the ring. If you leave one position open, setting hydrogen attachments on the other five, you will get a large number of molecules that have a non-halogen atom at that position. Therefore, most search systems allow you to use tools called generics to specify the substituents allowed at a particular position. There are generally two types of generics: system defined generics and user defined generics, or "R-groups."

5.2.3.1.1 System-defined Generics. The types of system-defined generics will vary widely from database to database. SciFinder® and the CAS Registry on STN® have a relatively small number of generics (called "variables"), while Reaxys® has a large number of extremely specific generics, inherited from the Beilstein system. However, there are several that are common to most systems.

X: any halogen
M: any metal
A: any atom except hydrogen
Q: any atom except carbon or hydrogen

In addition to these, most systems have variables that describe alkyl chains and cycles, occasionally being so granular as to include carbocycles and heterocycles as separate variables. Some more specialized databases include other, more specific variables. For example, the Cambridge Structural Database allows you to choose between a variable that represents any metal and one that represents any transition metal. It is always a good idea to consult the help file of your database to determine exactly what is available.

5.2.3.1.2 User-defined Generics. Often, the system-defined generics are far too general for your needs. You may wish to permit only chlorine, bromine, or iodine to be attached to your benzene ring, not astatine or fluorine. In this case, you need to devise your own generic group. Depending on the database, you can do this in one of two ways: by constructing an atom list or by making an R-group.

Atom Lists

An atom list is just what its name implies, a user-defined list of atoms that could be attached to a particular position in the substructure. They are most frequently employed when a searcher wants to vary metals permitted in a coordination compound, broaden the selection of heteroatoms at a given position in an organic molecule, or limit the type of attachments that can exist at a particular place in any structure. Reaxys®, the CAS Registry on STN®, and the Cambridge Structural Database are three systems that have atom list capabilities. Reaxys® and STN® also allow you to construct "not lists," or lists of atoms that may *not* appear at a particular position in your structure.

R-Groups

R-groups are more complicated, in that many systems allow you to construct them using fragments larger than a single atom. You need to look closely at the help file of your database for instructions on constructing an R-group because this technique varies widely from tool to tool. For example, at the time of this writing, SciFinder® does not permit users to draw their own fragments as part of R-groups, while Reaxys® and STN do. SciFinder® and STN® allow users to select R-group components from lists of "shortcuts" (shorthand notation indicating common organic groups), variables, and atoms, but Reaxys requires users to draw all R-group components on the screen. Since constructing an R-group by selecting pieces from a list is relatively simple, we will focus our attention on building an R-group using user-drawn fragments. In the course of this explanation, we will call the main portion of the substructure the "parent" and the parts that may vary "fragments."

5.2.3.1.3 Constructing R-Groups Using Fragments. There are four basic steps to constructing an R-group consisting of user-drawn fragments. The order in which one performs them varies depending on the database employed, but the help file of the tool should elucidate the exact procedure. One must draw the parent molecule, inserting a place-holder to represent the varying groups, and then draw the fragments that should attach at this position, being careful to assign them to the correct R-group. Finally, one must indicate the atom or atoms in the fragment that bond to atoms in the parent. The following example will illustrate all four steps.

Example 1 Using an R-Group to Vary Chain Substituents in a Structure

Assume that you wish to search for a 6,5 heterocyclic ring system with one of three different substitutents attached to the five membered ring. The parent ring system is shown in Figure 5.5.

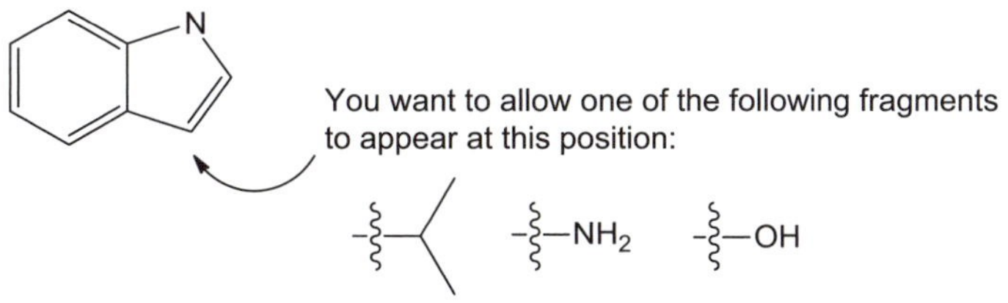

Figure 5.5 A 6,5 heterocyclic ring system that allows one of three different substituents at the atom indicated.

After you have drawn the main structure of the parent, attach a placeholder, in this case R_1, at the position to which the substituents should bond (Figure 5.6).

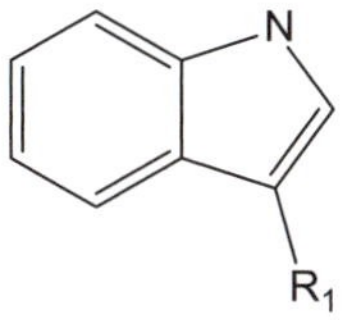

Figure 5.6 The 5,6 ring system with a placeholder drawn where the different substituents may attach to the parent.

Now, it is time to draw the substituents, or "fragments". Only draw the atoms actually present in the fragments themselves and not the bond that connects them to the parent (Figure 5.7).

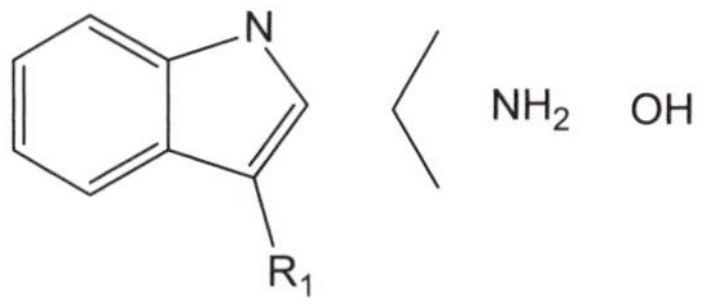

Figure 5.7 The 6,5 ring system with the fragments drawn to the right of the parent.

Next, you need to associate the fragments with the R-group to which they belong. This is generally done by selecting them and then specifying the R-group that owns them. Note that some systems, like STN, will have you draw and define the fragments before inserting the placeholder, while others, like Reaxys®, don't care what order you choose.

When you look at the completed structure, you will see that the R placeholder in the parent stands for one of the atoms in each fragment. Your last step is to tell the system which atom or atoms in the fragments bond to the atom or atoms adjacent to the placeholder in the parent and assign them "attachment points". For best results in most systems, all fragments should have the same number of attachment points, and the number of attachment points in each fragment must equal the number of bonds between atoms in the parent and the placeholder (Figure 5.8).

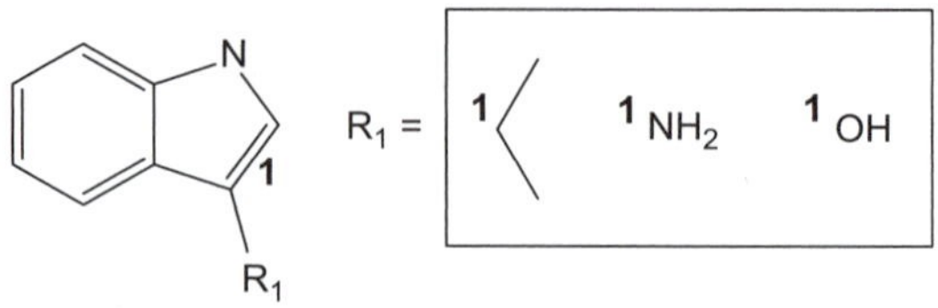

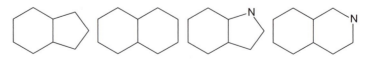

Figure 5.8 The completed substructure. The fragments have been grouped and assigned to the R₁ placeholder, and a point of attachment has been set on each fragment in the R-group.

When there is one bond between a single atom in the parent and one atom in each fragment, this finishes the task. However, if there are multiple bonds between parent and fragment, you need to pay attention to the orientation of the fragments. Example 2 will demonstrate this technique.

R-groups can also be used to perform a Boolean OR search for multiple substances that share a similar framework. To do this, one must first identify the atoms in each molecule that are common to all and use them to construct a parent. The atoms that differ become the fragments, as demonstrated in Example 2.

Example 2   Using an R-Group to Vary or Enlarge a Ring

Assume that you wish to use a single search to locate molecules that have one of the following four cores, shown in Figure 5.9.

Figure 5.9 Four 5,6 ring systems that can be located using a single substructure.

The first step is to identify the atoms that all four molecules have in common (highlighted in Figure 5.10). This will become the parent.

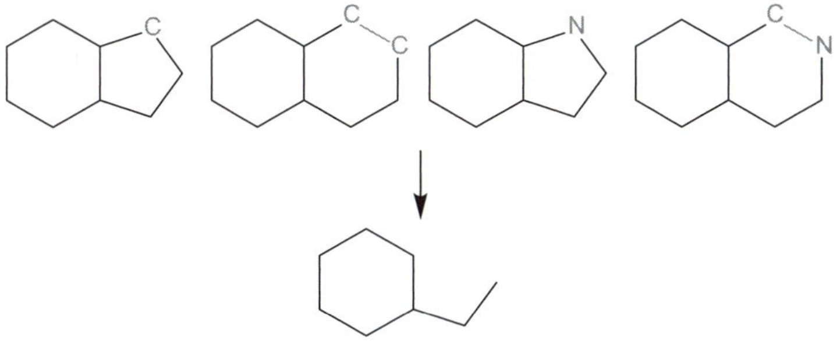

Figure 5.10   The portion of the four cores that all have in common is highlighted in black. This can be used to generate the parent, below.

Now, examine the parts of the molecule that vary (highlighted in Figure 5.11). These will become the fragments.

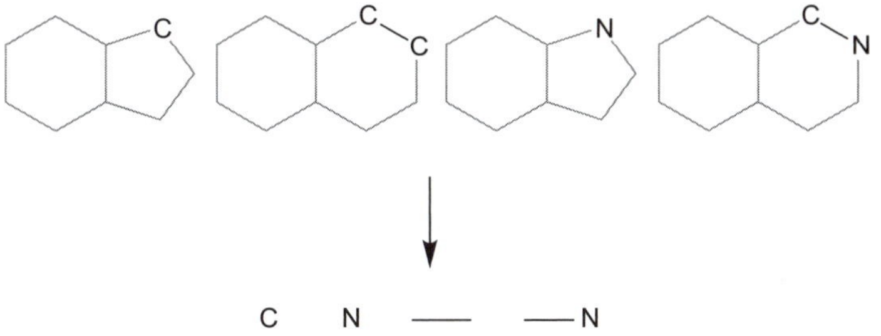

Figure 5.11   The portions of the four cores that differ are highlighted in black. These become the fragments in the search.

Insert a placeholder in the parent to represent the location of the fragments. In this case, since the varying segments are inside the ring, the placeholder must also be embedded in the ring (Figure 5.12).

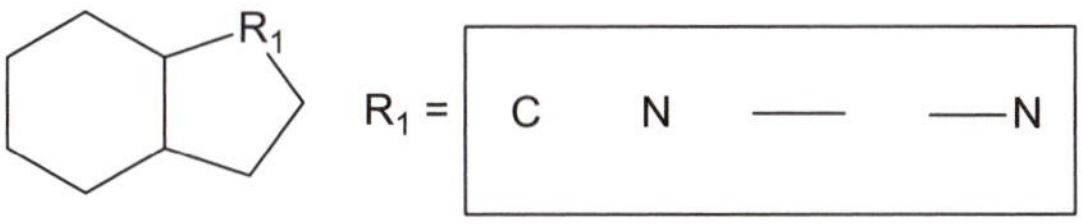

Figure 5.12   The substructure that will locate all four of the cores in Figure 5.9. The R_1 placeholder has been inserted into the parent, and the fragments have been assigned to R_1.

Finally, it is time to designate atoms in the fragments as attachment points. The placeholder forms two bonds to other atoms in the parent, so, each fragment will require two attachment points. Two of the fragments are single atoms, but this is fine; you simply put both attachment points on the same atom. The C-C fragment is symmetric, so, you can simply put one attachment point on one carbon and one on the other carbon. However, the C-N fragment is asymmetric, as is the parent, so, you will need to be careful to get the orientation correct when assigning attachment points. Methods of doing this vary from database to database, so, be sure to consult the help file of your chosen tool for more specifics. Bear in mind that most tools make you choose one unique orientation for each fragments, so, if you wish to allow either option, you will need to draw the fragment twice. The finished query structure is shown in Figure 5.13.

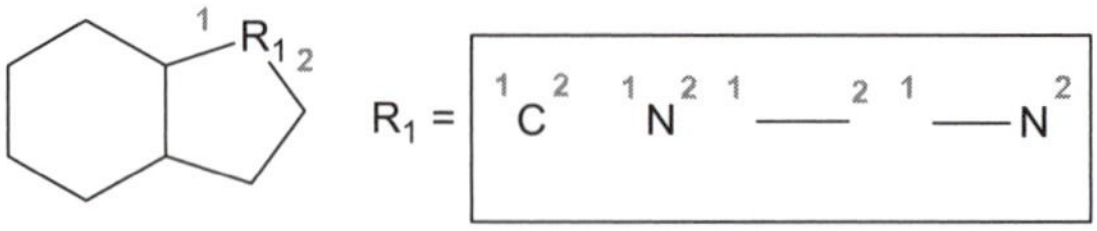

Figure 5.13 The completed substructure. The numbers 1 and 2 in each fragment refer to the sites in the parent to which each fragment atom bonds.

5.2.3.2 Repeating Units. Using an R-group is one way of expanding a ring, but embedding a series of increasing chains of atoms into a ring can be time-consuming and cumbersome. Depending on the search system and the desired substructure, using a tool called the "repeating unit" can be an attractive alternate method. SciFinder® and STN® both offer this option, allowing a searcher to select an atom or group in the structure and specify that it repeats a certain number of times in a row. When drawn into the structure, it looks exactly like the repeating group notation frequently seen in the literature (Figure 5.14).

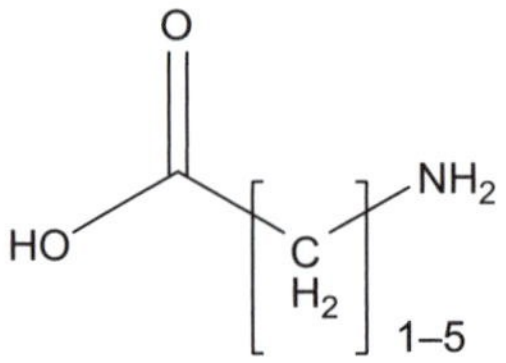

Figure 5.14 Common representation of a repeating unit. The CH_2 group appears anywhere from one to five times between the COOH and the NH_2.

The repeating unit technique has a few limitations, however. First, in SciFinder® and STN®, a unit may be repeated any number of times from zero to twenty, and it is not possible to try to bypass this limitation by placing two repeating units side by side. Second, a repeating unit may not be terminal; it must be placed between two other atoms in the structure. Finally, atom properties specified for any atom in the group to be repeated (limits on substitution, topology, *etc.*) will be repeated along with the atom.

At first glance, the ring-expansion query in Example 2 does not seem to be a good candidate for a repeating unit. However, if one remembers that a repeated segment may appear *zero* times in the molecule, in other words, it may not be present at all, an elegant solution quickly presents itself (Figure 5.15).

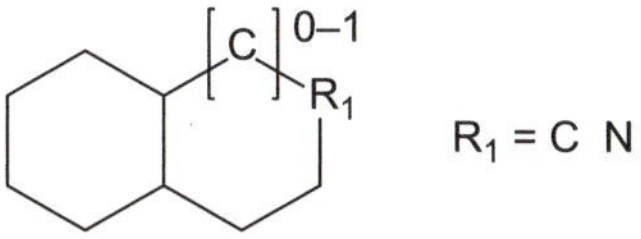

$R_1 = C \ N$

Figure 5.15 A method of locating the 5,6 ring system in Figure 5.9, employing a repeating unit.

When the repeated carbon appears zero times in the structure, the R-group is bonded directly to the six-membered ring, forming a 5,6 carbocyclic or heterocyclic system. If the repeated carbon appears once, the result is a 6,6 carbocyclic or heterocyclic system.

This technique can also be applied when searching for two slightly different cores, and it can help to get around SciFinder®'s lack of user-defined R-group functionality. For instance, assume that one wants to use a single search to find molecules with one of the two cores in Figure 5.16.

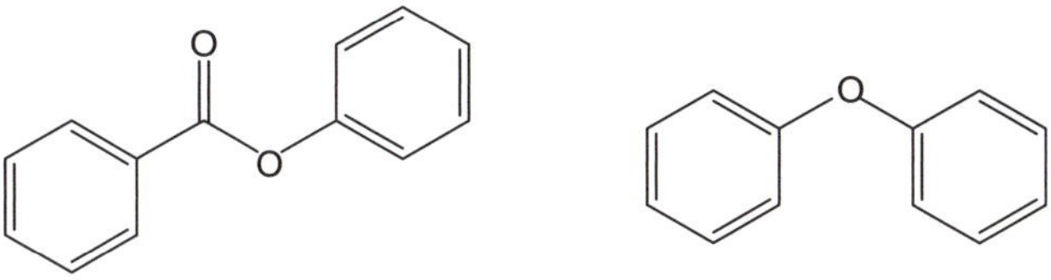

Figure 5.16 Two structures that could be located using either an R-group or a repeating unit.

One approach is to construct an R-group to tether the two rings (Figure 5.17a). However, this is not possible in SciFinder®, and, besides, it is easier to employ a repeating unit (Figure 5.17b).

5.2.3.3 Variable Points of Attachment. Suppose that a researcher wants to locate a single halogen attached to an asymmetric, carbocyclic core. He does not really care to which of the available carbon atoms in the ring the halogen attaches, but he is adamant that the

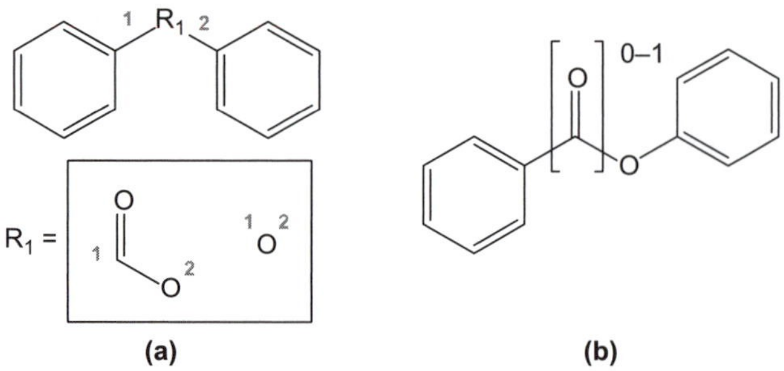

(a)   (b)

Figure 5.17   (a) An R-group that can be used to locate both of the structures in Figure 5.16. (b) A substructure that will locate the same two cores, this time employing a repeating unit.

ring remain otherwise unsubstituted. His first instinct is to use an R-group or an atom list, as portrayed in Figure 5.18.

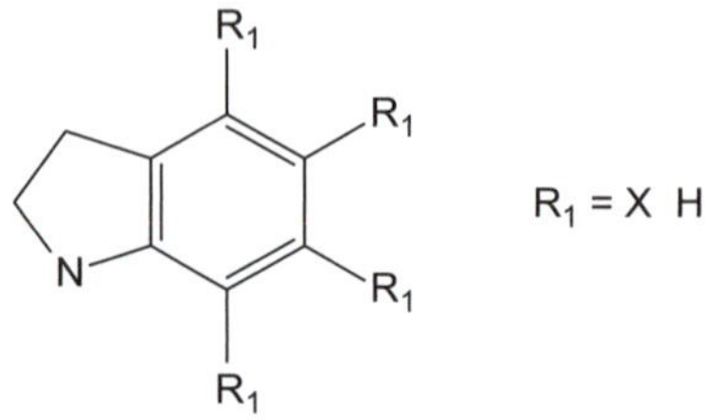

$R_1 = X$ H

Figure 5.18   An attempt to locate a monohalogenated ring system using a standard R-group.

However, this would allow any of the R placeholders to be either X or H, presenting the possibility of many levels of substitution: an unsubstituted ring, a fully halogenated ring, or any degree of substitution in between. While there are methods for eliminating the undesirable results from the hit set, these techniques involve many searches and a great deal of time. It is, therefore, much more effective to use a "variable point of attachment" (VPA) connection to attach the halogen to the ring. As with previous techniques, the availability of and procedure for using this type of connection differs based on the resource and structure editor employed. The procedure is very simple in SciFinder® and is demonstrated in Example 3.

Example 3   Using a Variable Point of Attachment to Halogenate a Ring

Assume that you wish to locate a 5,6 heterocyclic ring system, in which a single halogen attaches to one of the four open positions in the 6-membered ring, and the other three positions remain unsubstituted. First, draw the parent ring system, and then draw the variable X (representing any halogen) outside the ring (Figure 5.19a). Then, use the variable point of attachment tool to draw a connection from X to the four atoms in the parent, to which it may be connected (Figure 5.19b). Finally, use the Lock Atoms tool to lock out additional covalent bond formation on all four carbon atoms in the ring (Figure 5.19c).

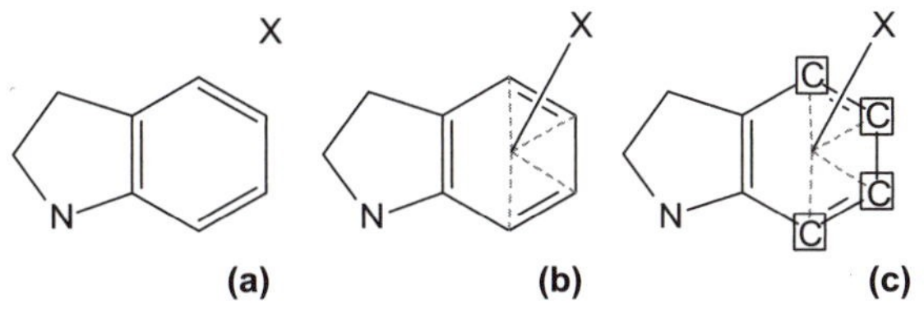

(a) (b) (c)

Figure 5.19 (a) The ring system and X variable are drawn into the structure editor. (b) Using the VPA tool, a bond is drawn from the X to each of the four unsubstituted carbon atoms in the six-membered ring. (c) The four carbons are locked to further non-hydrogen substitution using the "Lock Atoms" tool.

Because the X is already drawn in the structure, it is permitted to attach to any of the four positions to which it has been connected using the VPA, but the other three positions may only have H as a substituent.

While SciFinder® allows searchers to lock out non-hydrogen substitution, STN's® locking features will not accomplish this function in the same way. One can still locate a monosubstituted ring without specifying the location of the substituent, but the approach requires a bit more creativity. Instead of using a VPA for the substituent and locking all positions, one can use four VPAs: one connecting X to the parent and the other three connecting hydrogen atoms. The result is not the most elegant-looking of structures, but it accomplishes the desired function. The four variably attached atoms are interchangeable and will allow the searcher to retrieve all possible combinations (Figure 5.20).

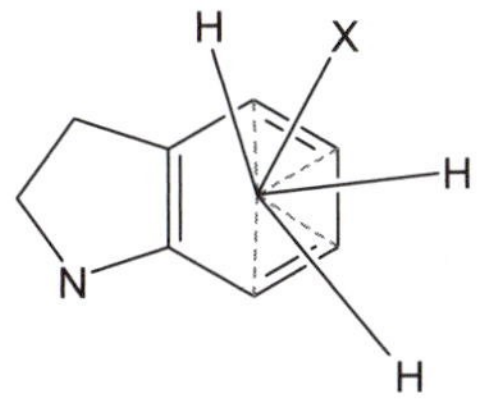

Figure 5.20 A method of locating a monohalogenated ring using four variable points of attachment attaching one X and three H atoms to the ring.

This technique can be adapted to retrieve bisubstituted rings using two variably-attached substituents and either locking out substitution or adding two variably-attached hydrogen atoms. One can also use a VPA to connect an R-group to a ring, giving a choice of substituent, as well as a choice of position.

Not all systems offer variable points of attachment, as such. For example, at the time of this writing, Reaxys uses a unique method for performing this task. It requires you to construct an R-group composed of the substituents to be variably attached and indicate the number of times that this R-group is permitted to appear in the structure. This is the only time that it makes sense to construct an R-group consisting of a single substituent; in other cases, it is better simply to draw the substituent into the parent. Example 4 gives more explicit instructions for constructing a VPA in Reaxys®.

Example 4   Constructing a Variable Point of Attachment in Reaxys®

Assume that you wish to use Reaxys® to locate a monohalogenated 5,6 heterocyclic ring system, where the halogen may appear anywhere on the six-membered ring. You must first construct an R-group consisting simply of the variable X (any halogen) and attach it to all four possible positions.

Right-clicking on the R-group definition allows you to access the R-group logic feature and set the maximum number of times that your R-group may appear in the structure, in this case once. Checking the box labeled "Others Hydrogen" ensures that the positions that do not contain the R-group will only be hydrogen-substituted (Figure 5.21).

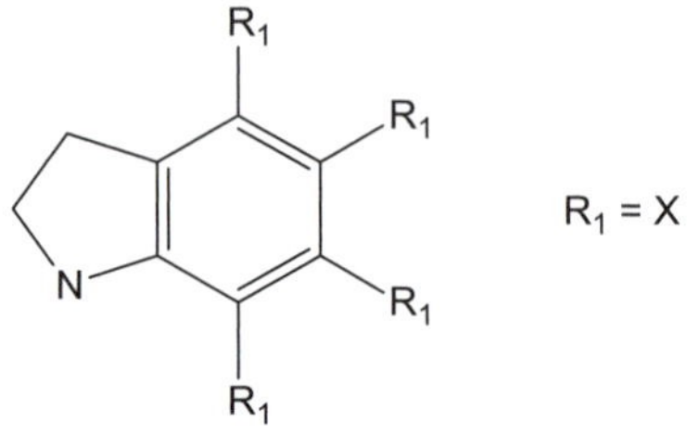

Figure 5.21   Reaxys' unique method of performing a VPA search. R_1 is defined to be X and is attached to every position that may be halogen-substituted.

5.2.3.4   Searching for Multiple Fragments in the Same Substance. R-groups and generics work very well when one knows exactly how

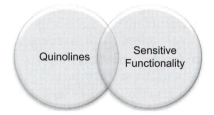

Figure 5.22   A Venn diagram illustrating the intersection between a set of substances containing the quinoline moiety and a particular sensitive functionality.

one wishes to connect one of several fragments to a particular part of a molecule, and a variable point of attachment allows a searcher to connect a known segment to one of several positions on a parent. However, neither of these techniques is of much use if one does not care how two fragments are connected in a molecule. Assume that a searcher is interested in making a quinoline-based substance. She believes that her synthesis will be complicated by the fact that her molecule also contains another, highly sensitive functional unit, and she would like to identify substances that contain both moieties to see if there are any reported syntheses of such products. In essence, she wishes to find the intersection between the set containing all quinolines and the set containing her other functional unit (Figure 5.22).

Depending on the tool chosen, there are several different ways of accomplishing this. Once again, it is extremely important to understand the capabilities of and features available in each resource.

Search for Both Segments Simultaneously

Many tools, including SciFinder® and Reaxys®, now allow a searcher to draw more than one fragment of a molecule into the structure editor without connecting them. When one enters more than one unconnected portion, the system generally searches for the fragments within the same molecule in a substance search and in different molecules in a reaction search, although Reaxys® does offer the option of searching for the fragments in separate molecules, as well. This is the easiest way to locate substances that contain two interesting substructures, but it is not terribly intuitive.

Refining by Substructure

Refining by substructure is another method of locating two substructures within the same molecule. Once again, both SciFinder® and Reaxys® permit this method of searching. It is best to begin with a search for the more complicated of the two structures. This also helps you to get a sense of the complete universe of substances that contain this primary substructure before limiting your results. Once you have retrieved all substances that contain the first substructure of interest, use the resource's "Refine by Substructure" option to input the second structure. When you submit the search, the system will filter your results to include only those molecules that contain the second fragment.

Combining Hit Sets

This is, by far, the most basic and least resource-specific method of performing the task at hand. You must first perform a search for substances that contain the primary segment and retrieve results. Next, repeat the procedure for the second segment. Finally, using the database's "combine hits" option (occasionally located in the "search history"), it is possible to combine the two sets using the Boolean operator AND. As usual, the procedure for combining hit sets varies from tool to tool; for example, at the time of this writing, SciFinder® requires users to save the first hit set before the "Combine Hits" feature becomes active. There are, however, major drawbacks to this technique, as compared to the previously-discussed methods. First, it is more time-consuming, requiring three separate searches to achieve a complete query. Second, it can occasionally be difficult to perform the initial two searches, particularly if one of the substructures is fairly general. If a search is too general, the resource will frequently fail to allow it to run to completion, making it difficult to create a complete, combined set of search results.

Combining hit sets can be used for other purposes than to search for a molecule containing multiple segments of interest. You may want to combine the hit sets from several different searches using a Boolean OR, thus creating a unified set of search results, which you can then either share or store easily or perform a common operation on all of the answers (finding certain physical properties of all represented substances, retrieving references including all of those, which will then be refined by text terms, *etc.*).

In some cases, you may perform a substructure search for a core of interest and retrieve many substances that have a common, undesired substructure. Again, there are two good ways of eliminating those undesired hits from the answer set. Some systems, like Reaxys®, will allow you to refine by substructure, draw the substructure that should be eliminated, and then opt to exclude substances from the answer set that contain this substructure. The second method, which is more universal but less elegant, requires you to perform separate searches for the desired and undesired substructures and combine the two hit sets using the Boolean NOT operator.

5.2.3.5 Setting Topology of Undrawn Substituents. It is relatively easy to set the topology of drawn substituents, but it takes a bit more creativity to restrict topological settings on atoms and connections that remain undrawn. In some cases, you may want to isolate a ring system or forbid additional ring fusion on one cycle in a system. At other times, you may wish to force a segment of a chain to have ring-only or chain-only topology. As usual, it is important to bear in mind that not every tool has the facilities to perform each operation described.

Isolating Rings or Ring Systems

There are two possible methods of isolating rings and ring systems from further ring fusion. The first is to forbid all undrawn rings. This option currently exists in Reaxys® and can be approximated in STN® through the use of a screen denoting the total number of rings in a structure. One should be aware that this will restrict rings to *only* those that have been drawn in the query structure, meaning that hit structures will include *no* additional rings, isolated or embedded. As a result, this method should be used with caution.

If forbidding additional rings in the structure is too draconian for your purposes, you also have the option of either locking ring fusion of a system or isolating a ring system. SciFinder® and STN® each have a method of indicating that a ring system should be isolated. This forces substituents on the ring to be chain-only but does not forbid additional rings in the structure. At most, it forces one chain bond to separate the isolated ring or system from the other rings in the structure. Bear in mind, when using this feature, that most search systems include spiro rings in their definition of fused rings, so isolating a ring system will also prevent spiro systems.

Forcing the Topology of Atom Attachments

This can be more challenging and resource-dependent than isolating a ring. For example, assume that you wish to locate the 5,6 ring system, shown in Figure 5.23. You wish to allow ring or chain substitution on

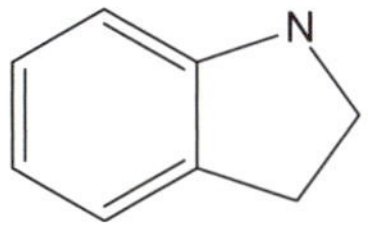

Figure 5.23   A 5,6 ring system, in which the topology of substituents attached to the six-membered ring is flexible, but which must have chain-only substituents attached to the five-membered ring.

the 6-membered ring, but you only want chains attached to the three open positions on the 5-membered ring.

STN® and Reaxys® will allow you to select each of the three carbon atoms in question and indicate that their attachments must be chain-only, either by setting the topology of non-hydrogen attachments (STN®) or by limiting the number of rings in which an atom may participate or ring bonds that it may form (Reaxys®). However, SciFinder does not allow this type of specification. Here, you will need to be a little more creative. There are a few possible approaches. If you wish to restrict the carbon atoms to sp3 hybridization, you can solve the problem using the structure in Figure 5.24, locking out ring formation on the bonds between R_1 and the ring.

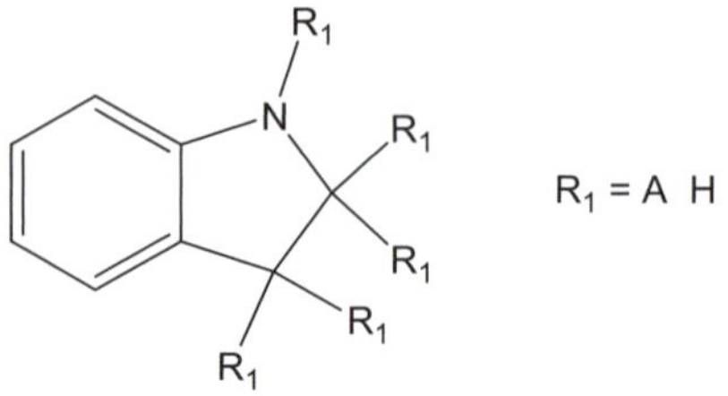

Figure 5.24   An attempt to force chain-only substitution on the five-membered ring. R_1 may be any atom on the periodic table, and the topology of the bonds between R_1 and the parent are set to chain-only topology.

If this is too restrictive, the best option is to search for the structure in Figure 5.24, as well as the two additional structures in Figure 5.25, combine all three searches with a Boolean OR, as described in Section 5.2.3.4.

It is also possible to use a variable bond order to connect a single R to each carbon, but this allows one free site to remain on each carbon that could, potentially, be filled by a ring attachment. Depending on the level of specificity desired of the results, this could be far too general.

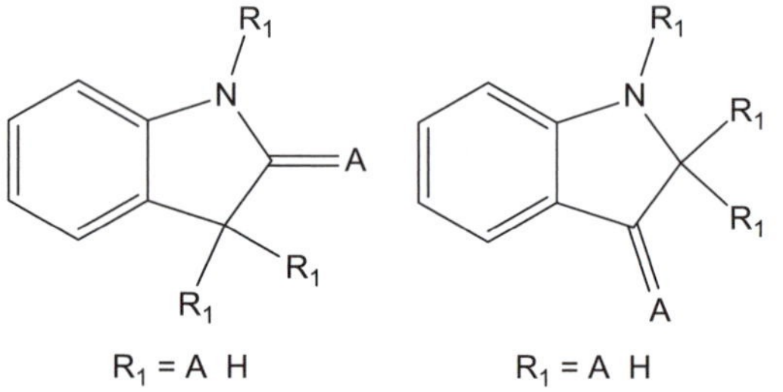

Figure 5.25   Two additional substructures that can be used to limit the topology of substituents on the five-membered ring to chain-only. These each allow one carbon on the ring to be sp2 hybridized. In both cases the topology of the bonds between R_1 and A are set to chain-only.

Using Ring-Only and Chain-Only Topology to Prohibit and Promote Ring Fusion

Assume that you are interested in locating substances containing the substituted 5,6 ring system in Figure 5.26. You do not care if the ring system is larger than the one drawn, nor do you care if the atoms in the chain participate in rings. However, you do not want the chain to cyclize with either the five- or six-membered ring in the system.

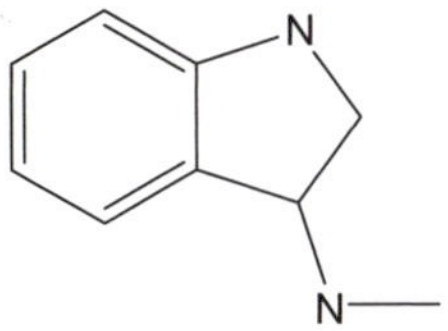

Figure 5.26   Substructure containing a 5,6 ring system and a side chain.

Depending on the search system you wish to use, it may not be necessary to isolate the ring system to accomplish your goals, nor must you lock the entire chain to ring fusion. Systems like Reaxys® and STN® let you set the topology of individual bonds in a chain. Therefore, you can define the bond between the nitrogen and the ring to be chain-only, allow the other bond in the chain to be ring or chain, and make no specifications about any other ring attachments. Forbidding topology on only the first bond of the chain leaves no possibility of cyclization with the ring system and accomplishes your goals. It is important to note that, as of this writing, SciFinder® does not allow a user to lock out ring formation on a segment of a chain, so this technique may not be used in that system. In SciFinder®, it is best to draw the substructure without the final carbon in the chain, lock ring formation between the nitrogen side chain and the ring, and analyze the resulting structures by atom attachment, selecting only those that have carbon attached to the nitrogen in the side chain. This will achieve the same results, but it is more time-consuming since it involves an additional step.

On the other hand, assume that you want the side chain to be part of the fused ring system, but you do not care about the size of the ring that contains it, the identity of the other atoms in the ring, or the site in the drawn 5,6 system at which it meets the rest of the molecule. If you assign the bonds in the chain ring-only topology, you will force them to fuse somewhere with the 5,6 ring system. As in the previous case, defining the topology on the first bond in the chain to be ring-only gives it no alternative but to cyclize with the rest of the substructure, given that spiro rings are defined as fused rings by most structure algorithms. Note that, in those systems that allow the exact topology of a single bond to be specified, ring topology may be added to any bond in a structure, not simply a bond adjacent to another ring. This technique may not be used in SciFinder®, which, at the time of this writing, has no way of forcing ring topology on anything that looks like a chain.

5.3 SEARCHING FOR ORGANOMETALLICS AND COORDINATION COMPOUNDS BY STRUCTURE AND SUBSTRUCTURE

Substructure searching for metal-containing species poses a different set of challenges than searching for organic substances. Many of these stem from fundamental inconsistencies of representation within databases and other search tools. Even the same source, such as *Gmelins Handbuch der Anorganischen Chemie*, searchable online *via* Reaxys®, may use a variety of methods of representing these types of substance. For example, the following zirconium complex could be represented in one of several different ways (Figure 5.27).

When dealing with metal-containing substances, therefore, it is crucial to consult the help files of the database. It can also be helpful to begin by performing a name or formula search for a known substance, so as to examine its structure and determine the conventions used by the resource. Many resources have begun adding tools that allow you to

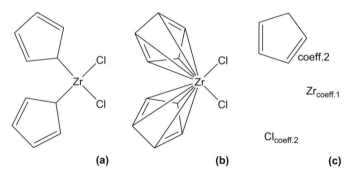

Figure 5.27   Three different methods of representing the same coordination compound (a) A single bond is drawn from each cyclopentadiene ring to the central zirconium. (b) A bond is drawn from every coordinating atom in the cyclopentadiene ring to the central zirconium. (c) The three components in the complex, cyclopentadiene, zirconium, and chlorine, are each drawn separately with a coefficient that indicates the number of each that appear in the complex.

generate a structure from a chemical's name, CAS Registry Number, SMILES string, or InChI, and this can also be an excellent way to get started. When using Reaxys®, it can also be helpful to use the ligand formula search option, described in Section 5.3.3.

5.3.1 Drawing Coordinations

Coordinations between metals and ligands are generally represented by drawing a connection between the metal and each atom in the ligand that coordinates to that metal. Most systems, such as SciFinder® and Science of Synthesis,[10] employ a plain, single bond for this purpose. However, the Cambridge Structural Database uses a connection called a "pi bond" to indicate metal-ligand coordinations. This more closely equates to what is happening chemically, but it can be hard to remember since it differs from most other search systems. Users of Reaxys® should employ a single bond to indicate metal-ligand coordinations, despite the fact that Reaxys®, structure editor includes a "coordination" in its menu of bond options. Reaxys® licenses a third-party drawing package called Marvin Sketch for structure editing, and, while it is customized to a certain extent, some features, like the "coordination," are strictly drawing tools and carry no significance in searching because they do not correlate to anything in the database's connection tables. If you are ever in any doubt about which type of connection to use, and the database help file does not provide adequate instruction, most search systems have an "any" bond option; as its name suggests, this connection will retrieve two atoms joined by any type of bond or coordination.

5.3.2 Defining the Number of Connections a Metal May Form

Setting the number of coordinations that a metal may form can be extremely challenging in any system. In many cases, one may use a method similar to defining the number of covalent bonds that a non-metal is permitted to make. Some systems, like SciFinder®, only allow a searcher to lock out coordinations that are not explicitly drawn in the structure, while other systems permit one to specify the exact number of coordinations or bonds formed by the metal, without drawing any of them.

In SciFinder®, you lock out coordinations on a metal in the same way that you lock out non-hydrogen substitution on a non-metal: using the "lock atoms" tool. However, after submitting a query incorporating a locked metal, searchers are frequently disconcerted to discover complexes in which the metal participates in additional bonds or coordinations. This is because the "lock atoms" tool's primary function is to lock out covalent bond formation, and metal coordinations fall into the "tautomerism" algorithm and are therefore permitted in the most general search. In order to force the metal to make no further connections, it is necessary both to lock the metal and to analyze the search results by precision, choosing only candidates that fall into the "conventional substructure" category. Those in the category of "closely associated tautomers and zwitterions" will include substances in which the metal may be further coordinated. Readers interested in learning more information about searching SciFinder® for such substances can refer to Ben Wagner's 2011 article on the subject.[11]

The major drawback to SciFinder® is that it is not possible to state exactly how many coordinations a metal may form without explicitly drawing them. Both Reaxys® and the Cambridge Structural Database allow you to select a metal and set the exact number of connections it forms. Reaxys® has one serious drawback in this regard though; at the time of this writing, there is no way to force an atom to form exactly six coordinations (S6). The database's search algorithm uses the S6 setting to represent maximal free sites, making it impossible to search only for octahedral substances, a common geometry for coordination compounds.

5.3.3 Ligand Formula Searching in Reaxys®

Given the challenges of substructure searching for coordination compounds, text-based searches for this class of substance can be very helpful. Elsevier's Reaxys® product has a novel method of performing text-based searches for such substances called "ligand formula searching." This technique focuses on the central metal atoms and the atoms in each ligand that coordinate to them, rather than looking at the precise or even general structures of the ligands. The ligand formula assigns a code to each ligand, based on its coordinating atoms, and lists these codes in alphabetical order following an alphabetical list of the metals in the complex. You are able to search using the complete ligand formula, called the "base formula," as well as the formulae and number of the individual ligands in the complex. An advanced searcher can even search for the ligands around a single metal center within a more complicated coordination compound, but this is beyond the scope of this chapter. We will deal with the construction of the base formula. For best results, we will exclusively use the "base formula" field within the "Multi-Center Ligand" category, regardless of the number of centers our complex has.

There are several steps to constructing a complete ligand formula.

List Metal Centers

Metal centers are listed in alphabetical order by their atomic symbols, and the symbols of all metals precede the codes for the ligands. Unfortunately, this gives no indication of the connections between the metals and the ligands. If more than one atom of a particular metal is present in the complex, the number of atoms of that metal follows its atomic symbol. For example, a complex containing two rhodium and one ruthenium centers would have its metal centers written like this: Rh2Ru.

Assign a Code to Each Ligand

There are two types of ligand that a complex may have, which we shall call "standard ligands" and "special ligands." Ligand formulae for standard ligands are constructed based on the types of atoms that coordinate to the metal centers, as well as the number of atoms that coordinate, also called the "denticity" of the ligand. Special ligands include carbonyl and cyano ligands, and they are assigned special codes that are not dependent on coordinating atom or denticity. We will first deal with the construction of a ligand formula for a standard ligand and then list the special ligands that do not follow these rules.

Examine the Types of Coordinating Atoms in Each Standard Ligand

Each coordinating atom in a ligand is given a code according to its placement on the periodic table. These codes are as follows:[12]

L = C
A = B, Si, Ge
D = N, P, As, Sb
Q = O, S, Se, Te
X = Halogens and hydrogen

For example, a ligand in which only nitrogen and oxygen coordinate to the metal would use the atom codes D and Q.

Determine the Denticity of Each Standard Ligand

As previously defined, denticity refers to the number of atoms in a ligand that coordinate to the metal or metals. The total number of atoms of each type that coordinate to one or more metals is indicated by putting the number in parentheses before the code; for example, a bidentate ligand in which two carbon atoms coordinate would be written as (2)L. As with the metals, a denticity of one is understood and does not need to be written; therefore, a halogen ligand is written simply as X, rather than as (1)X. The total denticity of the ligand must equal the sum of the number of atom codes in that ligand. Cyclopentadiene is represented as (5)L, showing a total denticity of five, with all coordinating atoms being L-type. Pyrrole, on the other hand, is represented as D(4)L, also showing a denticity of five but having one D-type and four L-type atoms that coordinate. Atom codes within a ligand are placed in alphabetical order.

Assign Codes to Special Ligands

Some ligands, such as carbonyls and cyano ligands, are very common. Rather than using L to represent these ligands, they are given special codes. There are six special ligand codes in Reaxys®. While three of them refer to specific structures, the other three are used to represent ligands containing particular elements, in which any of the elements could coordinate to the metal. These codes are[12]:

$CO = O \equiv C-$
$CS = S \equiv C-$
$CN = N \equiv C-$
$CNS = N \equiv C-S-$, $S=C=N-$, $N=C=S-$, *etc.*
$CNO = O=N \equiv C-$, $O=C=N-$, $N \equiv C-O-$, *etc.*
$CNR = $ Only C or N coordinates; R does not

Write the Completed Base Formula

The completed ligand formula begins with the metals in alphabetical order. Ligand formulae are enclosed in curly braces and are listed in alphabetical order, following the metals. Note that ligands that begin with a parenthetical number appear to take precedence over ligands beginning with a letter; when in doubt of the proper order, it is usually best to browse the database's ligand formula index. For more instructions on doing this, please refer to the database help files. Examining the ligand formulae of known complexes can also be helpful. If a complex contains more than one ligand with a particular ligand formula, the number of ligands with that formula is written after the close of the curly braces. For example, if two carbonyl ligands appear in a complex, their formula is written: {CO}2.

Example 5   Performing a Ligand Formula Search for a Complex with One Metal Center

Assume that you wish to find a complex similar to the one in Figure 5.28. This complex has a single metal center, zirconium, which is written as Zr. There are four ligands, two cyclopentadiene ligands and two chlorine ligands. The ligand formula for a cyclopentadiene ligand is (5)L, as has already been described, and the ligand formula for a halogen is X. Since there are two of each, each ligand formula is followed by the number 2: {(5)L}2 and {X}2. Finally, since 5 comes before X in alphabetical order, the formulae for the Cp ligands are listed first, followed by the halogens. The completed ligand formula for the complex is Zr{(5)L}2{X}2.

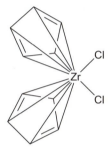

Figure 5.28   A zirconium complex, consisting of a single metal center coordinated to two cyclopentadiene rings and two chlorine atoms.

Example 6   Constructing a Ligand Formula Search for a Complex with Multiple Metal Centers

Now assume that you wish to find a complex with two cobalt centers and one chromium center, resembling the one in Figure 5.29. To construct the

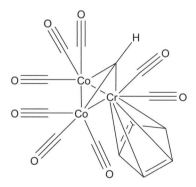

Figure 5.29   A complex coordination compound, consisting of three metal centers and ten ligands of various types.

base formula, first list the metals in alphabetical order, Co2Cr. *Next, assign a code to each ligand:* CO *for the carbonyls,* L *for the C-H, and* (5)L *for the cyclopentadiene. Finally, list the ligands in alphabetical order, followed by the number of each that appear in the complex. The completed base formula is* Co2Cr{(5)L}{CO}8{L}.

5.4   SIMILARITY SEARCHING

Substructure searching works quite well when you have specific requirements of the substances that one wishes to retrieve. However, if you are simply exploring the literature and looking for substances of interest, a similarity search can be even more interesting. To perform a similarity search, you need to draw an exact structure or a structure that includes R-groups or variables into the database's structure editor and search using the "similarity search" option, which is becoming more common in structure-based search engines. There are several different similarity search algorithms, including the Tanimoto algorithm,[13] which is used by SciFinder®.

Roughly put, a similarity search compares the structure drawn to all of the structures in the database and retrieves all of those that have a certain percentage similarity to the query structure. For example, SciFinder® bases its comparisons on similarities in atom count, ring count, bond sequence, elements, and type of ring fusion, among other things.[14] Resources that use similarity searching tend to rank search results from most similar (substances that are more than 99% identical to the structure drawn) to those that are least similar, and there is generally a threshold percent of similarity, below which results are not retrieved. This allows you to locate options that you may not have thought to include in a substructure search; for example, you may start with a substance that has a halogen leaving group and retrieve, though the similarity search, a similar substance with a completely different leaving group that might be superior under the conditions in which you are working.

5.5   A WORD ON MARKUSH SEARCHING

Chemical patents may contain specific structures of substances protected by the patent. These are indexed by secondary sources such as *Chemical Abstracts*; so, when one performs a substructure search using SciFinder® and STN®, one will retrieve some chemical patents. However, some patents include representations called Markush structures, which are used to claim a family of related compounds. They are named after Dr Eugene A. Markush, the first individual to successfully patent

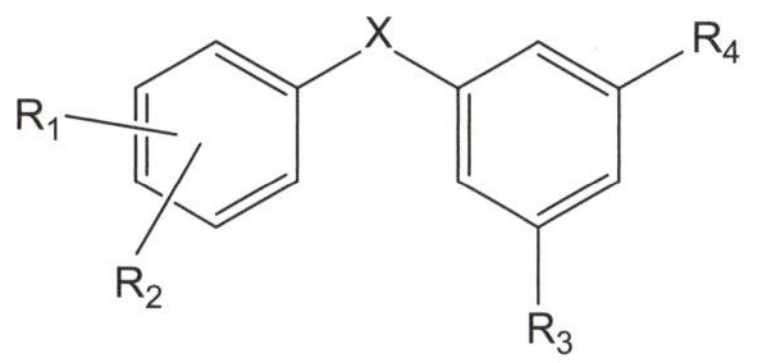

wherein

R_1 represents any electron withdrawing group

R_2 represents H, OH, COOH, COH, or COR', where R' is any alkyl, alkenyl, or alkynyl group

R_3 and R_4 represent any alkyl or cycloakyl group

X represents one of the following:

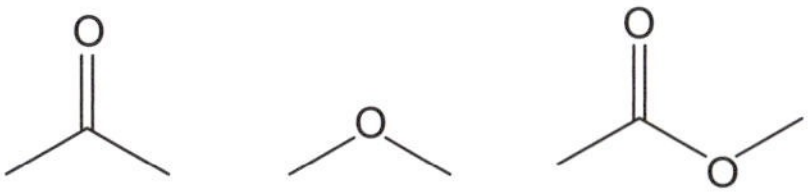

Figure 5.30 A fictitious Markush structure that could appear in the claims section of a chemical patent.

an invention including such a structure. A Markush structure, unlike other published structures, includes R-groups in the structural representation that appears in the patent, allowing a single structure to represent a wide variety of substances. For example, a Markush structure in a patent might look something like the structure in Figure 5.30.

Because the variables drawn can be extremely broadly defined (such as "any electron withdrawing group" or "any alkyl or cycloalkyl group"), Markush representations each protect thousands of different molecules, which may or may not actually have been synthesized in the lab. When you are performing an exhaustive search for all known substances with a particular substructure, you should supplement your search with a Markush search, as traditional substructure searching infrequently retrieves patents containing only Markush structures.

The two main databases used for Markush searching are MARPAT®,[15] from Chemical Abstracts Service, and the Merged Markush System (MMS),[16] from Questel Orbit. MMS is one of the few systems that has no graphical structure editor. Instead, searchers use a command-line interface to input strings of characters describing the atoms, connections, and variables allowed in the structure.[17] MARPAT®, on the other hand, is accessible through SciFinder® and STN® using the familiar structure tools in each system.

Although CAS has made Markush searching accessible to the end user by allowing users to perform quick and dirty searches of MARPAT® using the SciFinder® substructure search module, you should not attempt a comprehensive Markush search in SciFinder. STN® includes special Markush search options, which can be used to define the attributes of atoms and groups more specifically than SciFinder®'s algorithms allow. Because of the complexities of searching in MMS and MARPAT®, we recommend that scientists who wish to perform comprehensive structure searches through the patent literature consult a trained chemical information or patent information specialist, or, at the very least, contact the database providers for assistance or training materials.

5.6 CONCLUSION

Since chemistry papers frequently describe substances using their structures, chemical structure searching can be an excellent way of retrieving all information on a substance of interest. Structures can be relatively unambiguous, and substructure searching allows you to find many related substances using a single search, saving time and, in some cases, money. Substructure searching is like a puzzle; you must first think about all ways in which the database could possibly represent the desired substances and then construct a search that will retrieve all of them, while reducing the number of irrelevant hits. This creates a delicate balancing act; the more complicated the substructure, the longer it will take to devise and construct. In some cases, it is a better use of your time to perform a relatively simple search, and weed through results manually or use the database's refine features to whittle down the search set to a reasonably-sized group of relevant results. The best way to reach this balance is through experience. As you become more accustomed to the ways in which substances are represented in the various resources, it takes less time to construct a reasonable search. Finally, remember that the chemical information professional at your institution is always ready and willing to provide advice and assistance in search strategy and execution, and, if you do not have access to a chemical information professional, many of the database vendors provide complimentary or fee-based training and search assistance to their customers.

ACKNOWLEDGEMENTS

The author would like to thank Denise Callihan for her assistance and input, particularly her insights into Markush searching.

REFERENCES

1. W. A. Warr, *Wiley Interdiscip. Rev.: Comput. Mol. Sci.*, 2011, **1**, 557–579.
2. Sigma-Aldrich Catalog, http://www.sigmaaldrich.com/catalog/search/substructure/SubstructureSearchPage, Accessed July 23, 2012.
3. Chemical Abstracts Service, SciFinder®, https://scifinder.cas.org/, Accessed July 23, 2012.
4. Chemical Abstracts Service, STN® provides information you can trust, http://www.cas.org/products/stnfamily/index.html, Accessed July 23, 2012.
5. Reaxys®, https://www.reaxys.com/, Accessed July 23, 2012.
6. Chemical Abstracts Service, CAS History: Milestones, http://www.cas.org/aboutcas/cas100/annivhistory.html, Accessed July 20, 2012.
7. Chemical Abstracts Service, CAS REGISTRY(SM) and CAS Registry Numbers, http://www.cas.org/content/chemical-substances/faqs, Accessed July 20, 2012.
8. International Union of Pure and Applied Chemistry, Project Details: IUPAC – International Chemical Identifier, http://www.iupac.org/nc/home/projects/project-db/project-details.html?tx_wfqbe_pi1[project_nr]=2000-025-1-800, Accessed July 20, 2012.
9. Cambridge Structural Database, http://www.ccdc.cam.ac.uk/products/csd/, Accessed July 24, 2012.
10. Science of Synthesis, http://www.thieme-chemistry.com/en/products/reference-works/science-of-synthesis.html, Accessed August 26, 2013.
11. A. B. Wagner, *Issues in Science and Technology Librarianship*, 2011 DOI:10.5062/F4G44N6W.
12. Reaxys Help System, https://www.reaxys.com/reaxys/WebHelp/All_Files/Searching_Property_Data.htm, Accessed July 20, 2012.
13. T. T. Tanimoto, An elementary mathematical theory of classification and prediction, *International Business Machines Corp.*, New York, 1958.
14. Chemical Abstracts Service, Similarity Search Overview, http://www.cas.org/SCIFINDER/help/2007/SCH_Help/sf_only/tanimoto2.htm, Accessed August 27, 2013.
15. Chemical Abstracts Service, MARPAT® – The CAS Markush database containing the keys to generic substances in patents, http://www.cas.org/expertise/cascontent/marpat.html, Accessed July 23, 2012.
16. Questel, Merged Markush Service (MMS) – Chemistry, http://www.questel.com/prodsandservices/mms_chemistry.htm, Accessed July 23, 2012.
17. D. J. Bacino, unpublished work; D. Callihan, unpublished work.

CHAPTER 6

Physical Properties and Spectra

A. BEN WAGNER

University at Buffalo, Science and Engineering Information Center,
226 Capen Hall, Buffalo, NY 14260, USA
Email: abwagner@buffalo.edu

6.1 INTRODUCTION AND METHODOLOGY

Chemists often require physical properties and spectra for materials they are studying, yet finding that information can be a challenging task. Usually one desires specific properties for a specific substance measured under a particular set of conditions, such as temperature or pressure. With thousands of properties and millions of substances, the permutations are nearly endless even before one factors in a set of conditions. The searcher should begin with the realization that the particular data desired may have never been measured and, if measured, may never have been published. Even if published or otherwise available, the greater context of how data is produced and documented – why, how and by whom – can complicate both finding and using property data.

Numerous secondary data compilations exist that extract data values from the primary research literature. As helpful as these compilations are, the sheer number of them can actually create rather than remove barriers to finding needed information. Unlike most of the journal literature which now exists in electronic form, many key compilations are still available only in print. Other major data compilations are available electronically. Some are available for free on the Internet, while others

Chemical Information for Chemists: A Primer
Edited by Judith N. Currano and Dana L. Roth
© The Royal Society of Chemistry 2014
Published by the Royal Society of Chemistry, www.rsc.org

require costly annual subscriptions. Hence, access to the print and online resources of a major, long-established science library can greatly increase the chances of a successful search. Some of these data compilations are priced and marketed primarily to corporations under the assumption that the for-profit world will pay top prices for data required for commercial research and manufacturing.

This chapter focuses on the process of finding properties and particularly highlights the best free resources available to all. However, subscription resources often contain information not available freely on the Internet, may permit the searcher to find the desired data much more quickly, and can provide a higher degree of certainty that the property has never been published, should the search be unsuccessful.

To avoid long phrases, the term "properties" will usually be used as shorthand for the phrase "chemical/physical properties and spectra" throughout this chapter. Likewise, "compilations" will refer to all secondary sources including handbooks, databases, and indexes. Rather than providing lengthy lists and descriptions of the thousands of property sources, we will cover the best and most comprehensive sources. Additional resources such as web guides and books covering some of the older print literature in detail will be cited for those wishing to become expert searchers. Lack of mention of a particular resource or vendor should not be interpreted as a negative opinion by the author on their information content or utility.

There is a high degree of redundancy among the many secondary sources. A given measurement in a primary research article may be picked up in scores if not hundreds of print and electronic compilations and databases. As a result, a common property for a common substance may well appear in almost any source that one consults whereas an uncommon substance and/or an uncommon property may not appear in any data compilation, thereby requiring a search of the primary research literature. For common compounds, there often is extensive data in books and other old print sources such as the *International Critical Tables*,[1] for example, which, has detailed information on the density of sugar solutions.

Property data is often buried in very old literature as well as in gray literature such as patents, conference papers, technical reports, and dissertations. Hence, a careful searcher will consult databases and print sources that go back in time as far as possible and consult specialized databases such as the subscription *ProQuest Dissertations and Theses*[2] and the free *Espacenet Worldwide Patents*,[3] using carefully constructed keyword queries to ferret out literature references most likely to have property data embedded in the full text. One should not assume that the desired data is only in journal articles nor that older work is low quality and imprecise. Chapters 2 and 3 provide details on searching the primary literature, and Section 6.5.5 of this chapter briefly discusses strategies for finding properties embedded in the primary literature.

This chapter focuses on chemicals, *i.e.*, individual substances. Excluded were sources specific to toxicological and environmental information,[4–6] biomolecules such as enzymes and biosequences,[7–12] and property estimation software.[13,14]

Most of this chapter assumes one is starting with a compound or a set of compounds and wishes to find certain specific or all known properties for those substances. However, a number of the resources described here permit searching for ranges of properties values apart from specifying the substance in the query. The power of this technique, *e.g.*, Knovel's *Data Search* feature described in Section 6.4.2.3 or the *Combined Chemical Dictionary*,[15] a CRC Press database, is often underappreciated and seldom used. One can search for a set of desired property ranges and find substances suitable for whatever application the researcher has in mind without having any idea beforehand what type of materials would be a likely fit.

Not all property data is created equal. How properties are measured and documented impact the quality of the information one can find about the data. When using this data to build experiments and confirm results, it is critical that the data be accurate in what it is purporting to represent and reliable in the quality of the measurement process. Section 6.3 briefly discusses factors to consider in evaluating the quality and provenance of data found in the literature and on the web.

6.2 IDENTIFYING SUBSTANCES AND PROPERTIES FOR INFORMATION RETRIEVAL

Before consulting sources, the searcher should clearly understand the substance, the property, and the set of conditions desired.

6.2.1 Identifying Substances

For substances, the structure, molecular formula, synonyms, and class/ subclass should be determined. Print and electronic resources often have a number of different indexes or search fields, such as molecular formula or structure search capabilities. Especially with print indexes, the compound may appear in an index only under a single name. Above all, the proprietary, but widely used, Chemical Abstracts Service Registry Number (CASRN)[16] and the non-proprietary IUPAC International

Chemical Identifier (InChI)[17] should be identified. Both numbers are commonly used as unique identifiers in databases and handbooks.

The class and subclass should be clearly established. Given the many types of substances, most resources cover only certain classes such as metals, organometallics, polymers, or even more narrowly, alcohols or minerals. If necessary, a chemistry textbook can be consulted to establish the class.

It is also helpful at this stage to gauge how much is generally known about the desired compound. If volumes of information can readily be found on the Internet, it appears in common reference sources like the *Merck Index*, or thousands of hits are found by searching a major database like *SciFinder* (Section 6.4.2.1), then one is more likely to find unusual properties or a given property over a broader range of conditions such as temperature. Conversely, if a search of large systems such as *SciFinder* or *ChemSpider* turns up little or no information, it is unlikely the searcher will find more than a few of the simplest properties, or they may be completely unsuccessful.

6.2.1.1 ChemSpider. A wonderful new free resource, *ChemSpider*[18] not only provides all of the basic substance information itemized above, but it also is an excellent source of property information under the *Properties, Spectra, and Data Sources* headings. Maintained by the Royal Society of Chemistry, *ChemSpider* extracts a wide range of property and spectra information about chemicals by crawling the open Internet in much the same way that *Google Scholar* captures scholarly references.

6.2.1.2 Common Chemistry and Other Standard Sources. *Common Chemistry*,[19] created by the Chemical Abstracts Service (CAS), is an online chemical dictionary of about 7,900 common chemicals. It provides the CASRN, molecular formula, synonyms, structure, and Wikipedia link for each chemical. To augment such a resource, standard sources like the *Merck Index*, chemical supply catalogs, or classic print chemical dictionaries are often available in laboratory offices, library reference shelves, or online and are easy to consult.

6.2.1.3 CAS REGISTRY and Reaxys. For less common compounds, there is no substitute for access to the master online substance databases maintained by CAS (*File REGISTRY* on *STN International* or *SciFinder*) and by Elsevier's *Reaxys* system. CAS has identified over 65 million compounds to date, not counting biosequences. The substance information in *Reaxys* originated with the old *Beilstein and Gmelin Handbook* and now covers at least 15 million compounds. Both systems permit name, molecular formula, and structure searching. They also provide extensive property information and will be described in more detail in Sections 6.4.2.1 and 6.4.2.2.

The number of references and properties listed in the *CAS* files and *Reaxys* can be an excellent indication of the strategy the searcher should pursue, should the desired information not be directly found therein. For example, if there are over 10,000 literature references and hundreds of properties listed in these large databases, this indicates a commercial compound that will have a large variety of property information available in almost any resource one consults. A few thousand literature references usually indicate a compound of some research or environmental concern. A variety of properties are likely to be found. In the worst-case scenario, one finds only a small handful of properties or literature references. Frankly, it is unlikely that much more, if any, other information will be found even if a lengthy, exhaustive search of the numerous resources mentioned in this chapter is pursued. The compound simply has not been studied enough to be picked up in handbooks, compilations, and more focused databases.

6.2.2 Identifying Properties and Units

Information about the property can be found in a source a simple as *Wikipedia*, a textbook, or standard chemistry/physics handbooks. The International Union of Pure and Applied Chemistry (IUPAC) has two very helpful and authoritative resources:

- *Compendium of Chemical Terminology*, commonly known as the *Gold Book*.[20]
- *Quantities, Units and Symbols in Physical Chemistry* 2nd ed. (1993), commonly known as the *Green Book*.[21]

Note that only the older edition of the *Green Book* is freely available online. The 3rd edition (2007) must be purchased from RSC Publishing.

Questions that should be answered before proceeding with a search are:

- What are the commonly used units for the measurement or is it unitless?
- What variables change the value of the measurement, *e.g.*, temperature, pressure, or magnetic field strength?
- Are there synonyms, abbreviations, or related properties?
- Can this property be calculated from another property? (Be sure to also search for this other property.)

It is also helpful to determine the phase of the substance (gas, liquid, solid), given the conditions of interest. Some resources cover only specific phases.

6.3 EVALUATING PHYSICAL PROPERTIES SOURCES AND DATA PROVENANCE

Data provenance refers to the ability to trace and verify the creation of data and how it has been used, altered, or moved across different databases throughout its lifecycle.[22] This includes being able to evaluate the accuracy and precision of data as well as a clear definition of exactly what is being measured or reported. Data provenance can have a direct impact on scientific integrity, safety, and standardization. Scientific research is only as good as the data it is based on. The potential for meaningful contribution depends on the correlation of research results to other evidence and the revelation of novel observations.

It is critical that chemical property data be robust in what it is purporting to represent and reliable in the quality of the measurement process. Documentation, multiple sources of measurement, calibration of equipment, controls, and standard reference data are all factors that should be considered when evaluating the quality of physical property sources. They key questions that the user needs answers to are:

- Is the data reported actually measuring what was intended?
- Is it measuring what one needs to know?
- Was it executed in an accurate and precise enough manner for one's needs?

Evaluation of data quality examines how specific values were measured, reproducibility of results, consistency with other related properties and reports in the literature, and certified standards for the property.

The National Institute of Standards and Technology (NIST) has specified several levels of data evaluation to help scientists establish the reliability of property data. While the specific details of each level are far more involved than most chemists require, the categories, and how they are determined, provide a useful process to evaluate data for any purpose. The NIST Interactive Data Evaluation Assessment Tool[23] leads users through a series of basic questions concerning material specification, measurement method, correlations or models, source of data, and peer-review. It is simple to use and freely available. The questions and further information on the theory and practice of data evaluation for materials properties are detailed in the NIST Recommended Practice Guide Special Publication 960-11.[24]

In general, the following questions based on the NIST protocol should be asked about any chemical or physical property one finds in the literature or on the web. Most of this information should be available from the source of the data and cited in any further publications by the user of the data.

- Is the compound identified?
- Is the measurement method described?
- Was the data compared to certified reference values?
- Is the data consistent with other properties about the compound?
- Does the data agree with other independent measurements of the same property of the same compound?
- Is the source peer-reviewed or a commercial manufacturer?
- Is the source the original report or part of a secondary compilation?
- Is the data considered preliminary or included as supplemental information?
- If the source indicates unexpected results, does the research follow up and indicate methods to determine an explanation?

Although the provenance of data may not be completely described and the evaluation of data quality is seldom an easy task, it is still incumbent on a researcher to exercise due diligence in this area.

6.4 MAJOR ONLINE RESOURCES (THE PLACES TO START)

Assuming that one has clearly identified the substances and properties of interest and that, during that process, one has not already come across the desired information in the resources mentioned in Section 6.2.1, there are a number of prime free online databases that provide a wealth of property information. It is recommended that one starts here.

6.4.1 Best and Largest Free Compilations

6.4.1.1 NIST Chemistry WebBook and Data Gateway. The National Institute of Standards and Technology (NIST), formerly the National Bureau of Standards, is one of the premier creators, compilers, and evaluators of scientific data. For chemists, the foremost among their many databases is the free *NIST Chemistry WebBook*.[25] The *Chemistry WebBook* provides carefully curated property data on a large number of organic compounds plus a few common inorganics. Included are thermochemical, thermodynamic, gas phase ion, and spectral data.

Beyond the *Chemistry WebBook*, NIST publishes a large number of additional databases, some free and some that must be purchased. The *NIST Data Gateway* is a convenient metasearch engine across most of them.[26] The Gateway permits searching by specific keywords, properties, and substances/materials. A chemical kinetics and a biofuels database are just two examples of the rich information available from NIST. NIST has also reissued much of the thermophysical data originally published in print by the Thermodynamics Research Center[27] (Section 6.7.3.1).

6.4.1.2 ChemSpider. The previously mentioned *ChemSpider* (Section 6.2.1.1), by mining the open web, is a tremendous time saver. Generally only reliable sources, nearly always explicitly referenced, are captured by their web crawlers. As currently structured, the *Properties* heading provides basic property and safety information, much of it taken from material safety data sheets. Tabs also provide predicted properties. The *Data Sources* heading displays a series of tabs providing a rich set of physical, biological, metabolic, toxicology, and environmental data in addition to more spectral and safety data. There is also a separate *Spectra* Heading.

6.4.1.3 MatWeb. Although not strictly chemical, but certainly of interest to chemists, *MatWeb*[28] is an unusually comprehensive commercial materials web site, covering over 88,000 materials. Included are polymers, metals, alloys, ceramics, glass, fibers, composites, semiconductors, wood, stone, and other engineering materials. Up to approximately 75 properties are available for each material. Searching can be done by material type, trade name, specification number, manufacturer, up to three different property ranges, and alloy composition ranges. *MatWeb* also encourages submission of corrections and additional materials and data from companies. A few features are available only to those that register for free or pay a subscription fee to become premium users. For example, premium users can download data in various formats, simultaneously search up to 10 criteria from the Advanced Search page, and sort results based on property value.

6.4.1.4 ChemExper Chemical Directory. ChemExper Chemical Directory[29] is a metasearch of over 9 million chemicals from over 2,300 suppliers. What sets this site apart is the unusually robust search engine allowing queries by CAS Registry Number, InChI identifier, molecular formula and weight, name, substructure, and a number of different property values and ranges. Only the basic properties are directly provided, *e.g.*, density, melting point, boiling point, and refractive index. However, linked suppler web sites and material safety data sheets will usually provide some additional properties. As is often the case with manufacturer and supplier information, the property data may not be critically reviewed, requiring a degree of caution. This site continues to add enhanced features such as an impressive free 3D molecular viewer.

In singling out these few top resources, the author realizes there are many other excellent free sources of property information, small and large. Ultimately, the best resource is the one that has the property one is seeking. However, it is the judgment of the author that the free resources itemized above are the best places to start most property searches.

6.4.2 Best and Largest Subscription Compilations

As helpful as the free resources covered in the previous section can be, it must be stressed that ultimately there is no substitute for access to large subscription databases described in this section. Depending on the size and nature of one's institution, the cost of these subscriptions can be moderately to very expensive. If one's organization does not subscribe to these resources and one lives within convenient traveling distance to a major university library, these resources typically can be accessed within the library facility (but not off-campus) by walk-in patrons. Some electronic resources have licenses prohibiting commercial use, limiting downloads of citations, and other restrictions. The library staff will be able to explain what is available and under what conditions.

Finally, most vendors and publishers will happily arrange trial access for organizations, allowing for thorough testing before making any purchase decision.

6.4.2.1 SciFinder/Chemical Abstracts Databases. Chemical Abstracts (File *CAplus*)[30] and related databases produced by Chemical Abstracts Service (CAS) are available on a number of pay-as-you-go and flat-rate search systems such as *ProQuest Dialog*,[31] *SciFinder*,[32] and *STN International*,[33] the last two mentioned are associated with CAS. Since 1907, CAS has abstracted the chemical literature, indexing both substances and concepts in millions of published documents.

Although the text of some abstracts may directly report a few properties, until recently, one typically had to consult the full text of articles to find the actual property values. However, in 2002, CAS started adding experimental data from both internal and externally licensed sources displayed in a table that is part of the detailed substance record in *SciFinder/File REGISTRY*.[34] The first licensed source was the *SPRESI* database[35] produced by ZIC/VINITI. The large Wiley spectra collections have now been licensed (Section 6.7.2.2). Each property value and spectra are cross referenced to the original source. CAS has provided an online document[36] detailing the sources for their experimental property and spectra, some of which are digitized.

In addition, since 2001, CAS has added calculated properties and NMR spectra for most organic compounds, using Advanced Chemistry Development (ACD/Labs) software.[37]

In SciFinder, the "Explore References" search query screen automatically searches both *CAplus* (1907+) and *MEDLINE*[38] (1946+) literature references. From 1985 on, some *MEDLINE* records include CAS Registry Numbers.

It is beyond the scope of this chapter to provide detailed instructions for each of the many implementations of the CAS files. The author recommends that searchers consult the vendor training materials, including webinars, for the particular system they access. However, the calculated and experimental properties are easily found in most systems by displaying the detailed substance record. Most systems also permit searching for a specific value or a range of values for certain properties.

Although millions of properties are now provided directly in *File REGISTRY*, this is still a fairly small fraction of property data reported in the primary literature indexed in *CAplus*. Hence, if the information desired is not in the property table associated with the substance record, one should do a keyword search for the desired property, in combination with the registry number(s) of the compound or the compounds class. For example, a *SciFinder* query "dielectric constants of epoxy resins" would retrieve many appropriate references. *SciFinder* uses a natural language query system and automatically truncates search terms at appropriate points in order to retrieve variant word forms. All implementations of CAS files permit starting with a group of substances records and crossing them over to retrieve all literature references, which can be further refined by property keywords.

6.4.2.2 Reaxys. *Reaxys*[39] from Elsevier is a new system with some very old data dating back to 1771. Included is information from two large, classic print handbooks originally produced by the Beilstein and Gmelin Institutes. These German "handbuchs" masterfully extracted chemical and physical property data and systematically arranged all this information by compound class. In most cases, the actual data values were recorded, often making it unnecessary to retrieve the original article. More details on the history, scope, and relationship between the print and online versions of Beilstein and Gmelin are provided in Section 6.5.3, which focuses on the print handbooks.

The inorganic and organometallic portion (Gmelin) is currently compiled and owned by Gesellschaft Deutscher Chemiker (GDCh) and licensed exclusively to Elsevier. The organics portion (originally Beilstein) is now produced and exclusively owned by Elsevier (the Beilstein Institute is no longer associated with the data that for so long was produced by and bore its name). Elsevier has also added a patent database and markets the entire *Reaxys* set of databases as a bundle.

The data now in *Reaxys* represents the largest set of carefully reviewed property data in the world, easily millions of values for at least 140 different chemical, physical, and biological properties. On the "Substances & Properties" search query screen, one can search for any combination of properties using exact or a range of values in combination with a substructure, if desired. A searcher has the option of a simple form-based or an advanced property search panel. Using the advanced option, one has full use of Boolean operators to perform searches using any combination of properties, substructure, bibliographic, and reaction information. One can also do a search for keywords across all text fields. Many systems limit direct searching of properties to a small subset of the total number of properties available for display. *Reaxys* is unusual in that every property field is directly searchable. Substance and related property information can be downloaded into a spreadsheet.

To give some sense of the amount of information available, currently the substance, benzotrifluoride (CASRN 98-08-8), has 286 property values, 150 spectra, and 543 reactions all drawn from 763 references. Any set of fields or the entire record can be viewed with the information presented in a very readable tabular format.

The *Reaxys* Training and Support site[40] contains brochures, presentations, and links to webinars that clearly demonstrate the power of this versatile search system.

6.4.2.3 Knovel. Fairly new to the database scene, *Knovel*[41] has built up a large full-text collection of over 4,000 handbooks from many different publishers in over 25 technical areas matched with an exceptional search system that can search down to the level of individual cells in data tables. Many of the books are enhanced with features such as interactive tables and graphs that allow you to manipulate, customize, and download the data. Often, the handbooks also provide extended textual descriptions that are rich in background information and technical details.

Subject coverage includes: adhesives, aerospace technologies, biochemistry, biology, biotechnology, ceramics, chemistry, chemical engineering, construction materials and engineering, electrical and power engineering, environmental engineering, food science, mechanics and mechanical engineering, metals and metallurgy, pharmaceuticals, plastics and rubbers, safety, health and hygiene, and semiconductors and electronics. Access to a number of "premium" handbooks and sets of handbooks requires an additional subscription fee.

In addition to the default simple keyword search box, *Knovel* provides a full-feature Data Search query screen that permits searching materials in combination with exact or range searching on multiple properties. Unfortunately, a few publishers such as McGraw-Hill have recently withdrawn their content from *Knovel*. However, *Knovel* remains a first choice for many reference and data questions among librarians fortunate enough to have access to the database.

Knovel provides a small number of useful databases to non-subscribers (free registration required). Of special note for property searching are the *International Critical Tables*[42] and *Knovel Critical Tables*.[43] Each week, registered non-subscribers also get 12 days of free access to two subscription titles chosen by *Knovel* from their collection of nearly 2,000 handbooks, databases, and conference proceedings.

Knovel is a good example of an online book platform whose search system and extensive content transforms it into a very useful reference database. From the beginning, it has made properties data sources a primary collection focus.

6.4.2.4 SpringerMaterials. *SpringerMaterials*[44] is another recently introduced online database based on a massive, classic, print reference work, *Zahlenwerte und Funktionen aus Physik, Chemie, Astronomie, Geophysik und Technik* commonly known as *Landolt Börnstein*, named for the two original editors in 1883. The original 6th edition volumes, approximately 50 in number, are in German. The mostly English New Series, which started publication in 1961 and currently exceeds 400 volumes, is included in *SpringerMaterials*. The print version is more fully described in Section 6.5.4.

Suffice it to say, *Landolt Börnstein* is one of the largest and most carefully curated property data collections in the world. The publisher states that *SpringerMaterials* covers 250,000 substances and material systems, 3,000 properties, and 1,200,000 literature references to date. The database is updated quarterly and includes additional supplemental databases such as the *Linus Pauling Files*, a comprehensive database on inorganic solid phases, and the thermophysical properties subset of the *Dortmund Data Bank Software & Separation Technology* (DDBST).

It is almost impossible to overstate the breadth of this resource covering from Abbe's Number to nebulae to zone melting and the proverbial everything in between. The table of contents/metadata is freely searchable, making it easy to identify if the database has the desired information. As with Google Books, a short snippet of full text is provided. Using the Advanced Search tab, no subscription is required to search the entire database by keywords, bibliographic reference, chemical name, molecular formula, CASRN, and property name. The chemical substance record, also freely available, lists synonyms and provides a stick-and-ball 3D image that can be rotated by the user. The locations of all tables within *Landolt Börnstein* that contain information about the compound are itemized. Only the full text record requires a subscription. A number of major university libraries maintain useful guides to both the online and print versions, including the California Institute of Technology.[45]

6.4.2.5 Society and Research Center Compilations. As might be expected, societies and various private, governmental, and academic research centers have often created and/or complied significant data and spectra collections of interest to their clients or members, funded by external agencies or simply developed as a normal part of their research effort. A searcher will find it worth considering what organizations would have an interest in or a need to develop the particular property data of interest. For example, this approach could be as easy as visiting the publications or resources section of polymer society web sites if properties of commercial polymers are desired.

These compilations are too numerous to itemize and are available in almost any format imaginable from very old print tomes to cutting-edge web sites. However, three efforts deserve particular mention, though it again should be stressed there are many other excellent organizations producing good data.

The Design Institute of Physical Properties (DIPPR)[46] of the American Institute for Chemical Engineers (AIChE) has developed a set of databases and publications that provide detailed, high quality property data for significant commercial chemicals. For example, their *Thermophysical Properties – DIPPR Project 801* covers over 2,000 compounds, listing values for 49 thermophysical, 34 constant, and 15 temperature-dependent properties.

The *DIPPR Project ESP* (Environmental and Safety Properties) database contains critically evaluated properties of regulated chemicals and other chemicals of interest to DIPPR sponsors. The database is intended to support engineering and regulatory calculations and, when used in conjunction with its estimation protocols, to predict properties not readily available from the literature.

DIPPR is now a licensed distributor of the *DETHERM* database in the Americas. *DETHERM* provides thermophysical property data oriented toward construction and design of chemical apparatus and plant processes. It is produced by DECHEMA e.V. (Gesellschaft für Chemische Technik und Biotechnologie or Society for Chemical Engineering and Biotechnology)[47] in cooperation with DDBST GmbH, Oldenburg, and Fiz Chemie in Berlin.

The database covers about 36,500 pure compounds and 124,000 mixtures and also includes literature references. Phase equilibrium, vapor pressure, critical data, thermodynamic, transport, surface tension, and electrolyte properties are available *via* an in-house or web database. The online database can be searched for free to determine if the desired data is available, but display of the actual data and literature references requires a subscription. The database can also be accessed on the STN International search system.

CINDAS LLC[48] is a private company formed to exclusively disseminate materials properties data collected and analyzed by the Center for Information and Numerical Data Analysis and Synthesis (CINDAS) at Purdue University. Their product catalog[49] page lists five databases including the *Microelectronics Packaging Materials Database*, the *Structural Alloys Handbook*, and the *Thermophysical Properties of Matter Database*.

6.5 USING HANDBOOKS TO FIND PROPERTIES AND PROFILE SUBSTANCES

Now that many publishers and aggregators offer large collections of handbooks electronically, the distinction between handbooks, electronic reference books, and databases is becoming increasingly blurred. Examples from individual publishers include the *CRCnetBASE* products from Taylor and Francis and the new *AccessEngineering* from McGraw-Hill. *Knovel*, previously discussed in Section 6.4.2.3, is one of the foremost aggregators of handbooks from a variety of publishers.

The term "handbook" is used to describe a variety of materials. Some handbooks contain introductory or advanced textual descriptions while others are almost exclusively filled with data tables. They range from compact single-volume works to large systematic reference works published over many decades in hundreds of volumes. Obviously, we will focus on the data-oriented handbooks, though even those number in the thousands. Hence, only a sampling of the most prominent ones can be discussed in this section.

6.5.1   How to Find and Use Print Handbooks

Despite the thousands of handbooks available in databases or stand-alone online books, many valuable, classic property handbooks have not been digitized. Even if an electronic version is available, few institutions can afford to subscribe to all the property resources of interest to their patrons, especially when often a significant investment in print versions has already been made. Though the reliance on instantly accessible and easily searched electronic resources is understandable, a good searcher will not ignore print volumes sitting on increasingly little-used reference shelves in the library or information center. Although many advances have been made in instrumentation and methodology, searchers should not assume that older measurements are necessarily imprecise or poorly done. In a word, this is engaging in chronological snobbery, which is a pitfall for any searcher.

There are two basic approaches to identifying print handbooks:

1. Publicly available online library catalogs.
2. Literature guides, often created and carefully maintained by science librarians.

Online guides will be discussed more fully in Section 6.6. Although online guides are useful in finding both historical and current handbooks and other property resources, a hard copy guide book to the chemical literature can provide important, additional information about classic and discontinued resources.[50,51]

Despite the numerous search interfaces for library catalogs, nearly all have options for keyword, "title begins with", and title word queries. When looking for reference material, keyword searching may return an overabundance of results. In those cases, one should limit the search to title words, see if there is a way to limit the keyword search to the reference collection, or use other limit or refine options. The searcher may need to switch to an "advanced search" query screen to have some of these options available. Using the catalog at the author's university as an example, a search for "thermodynamic*" gave over 2,000 hits as a keyword and half as many as a title word, of which 37 items were on the reference shelves. Nearly every one of the 37 reference works were packed with thermodynamic data. Other good strategies are to browse Library of Congress Subject Headings that use the standard subheading

"Handbooks, Manuals, *etc.*" or to include the keyword "properties" in the query.

Once an appropriate print handbook has been located, it is important to study the front matter (preface, *etc.*) and back matter (indexes, notes, *etc.*), not only the entire book, but also the explanatory matter before and after each table, section, and chapter. Particular attention needs to be paid to the basic arrangement of the book, *e.g.*, by name or molecular formula, and available indexes, *e.g.*, CAS registry numbers or chemical classes. Are the chemical names in the basic arrangement of the volume or indexes inverted (phenol, 4-chloro-) or direct (4-chlorophenol)? Such advice may seem obvious to some searchers, but the simple art of understanding how to approach a print resource is in danger of being lost in our increasingly electronic world.

Print reference sources always explain abbreviations, arrangement of materials, coding systems, and sources of the data, if the user will only take the time required to read about the explanatory information before attempting to dive into the middle of the handbook to find something. Table column headings may use symbols that are explained in a key to the table which often appears only on the first page of the table even though the table may span many pages.

Library guides to physical property resources often contain a mixture of print, free Internet resources, and subscription databases available at that particular institution. A discussion of library guides is contained in Section 6.6.

6.5.2 Online Handbooks

6.5.2.1 CRC Handbook of Chemistry and Physics. This handbook[52] is the online edition of the venerable one-volume handbook published annually since 1913, a standard fixture in many chemical laboratories. Significant updates occur in every edition, including adding and removing entire tables with the goal of having the most useful, current data available within the limits of a single volume work. All data is critically reviewed. The online version contains an easily navigated table of contents and keyword, structure, CAS Registry Number (use double quotes around number), and property data searching. Along with much other information, physical and chemical properties of compounds, polymers, fluids, isotopes, and atomic particles are provided in well-documented tables. Range searching of properties is available.

The online version is provided as an annual subscription, part of the *CRCnetBASE* product line by Taylor and Francis described in the next section.

6.5.2.2 CRCNetBASE Products. In 2003, Taylor and Francis acquired CRC Press, which published many classic print handbooks. From that core, Taylor and Francis developed dozens of sets of bundled online handbooks accessible through a single federated search platform branded as *CRCnetBASE.*[53] In addition to standard collections, custom bundles can be subscribed to under the *netBASE Select* option. Environment, polymer, engineering, materials science, life science and many more bundles can be subscribed to. Unfortunately, the CRCnetBASE query screen only permits full text, title, author, and ISBN/DOI searching. Without a property data search screen, it can be a challenge to zero in on the data tables and graphs.

6.5.2.3 ASM Handbook. The *ASM Handbook,*[54] available both as a set of print volumes and as an online subscription database, is a good example of a valuable contribution from a society publisher. The handbook (and the society) originally covered only metals and alloys, but now has expanded into engineering materials, ceramics, composites, and structural plastics. Although full text access requires a subscription, the table of contents can be browsed and the contents can be searched with little snippets of text for each retrieved section displayed, much like *Google Books*. As might be expected, this source focuses on engineering properties like fatigue, corrosion, and mechanical strength. In addition, ASM licenses their content to certain platforms, including *Knovel*.

6.5.2.4 Tables of Physical and Chemical Constants. Thanks to the National Physical Laboratory (Middlesex, UK), the entire text of the 16th edition of Kaye and Laby's *Tables of Physical and Chemical Constants*[55] is available on the Internet for free. Included are all data tables, formulas, graphs, and charts. For those not able to afford online subscriptions to the databases discussed above, this provides a more limited, but useful alternative. Thermodynamic, electrical, mechanical, and acoustical data are just a few examples of the wealth of data provided. The table of contents is easily browsed, but searching is limited to keywords in the text.

6.5.3 The Beilstein and Gmelin Handbooks

No review of physical property sources would be complete without a discussion of the two classic resources compiled by two German institutes and bearing the names of their first two editors (Konrad Beilstein and Leopold Gmelin) with origins in the 19th century. The Beilstein

Handbook, started in 1881, covers organic compounds and a small number of organometallics (having Li, Na, K, Rb, Cs, Mg, Ca, Sr, or Ba as central atoms and carbides, cyanides, cyanates, and thiocyanates with carbon-metal bonds). The Gmelin Handbook, started in 1817, covers inorganics and the rest of the organometallics not in Beilstein. The only database system that contains all of the information that was digitized from these handbooks plus continuing electronic updates is *Reaxys*, which was discussed under major online subscription databases in Section 6.4.2.2.

As of this writing, there is some information in the printed Beilstein and, even more so, printed Gmelin books that does not appear in the online database, and *vice versa*. Some textual descriptions and all graphs and diagrams have not been digitized. However, most of the actual property data is available electronically. The major exception known to the author is that the printed Gmelin volumes and the online Gmelin information for 1976–1997 were apparently compiled completely independently of each other, with the online edition covering a smaller set of journal titles. Hence, for exhaustive searches, a researcher should consult both print volumes and Reaxys. In the early 1980s, both handbooks switched from German to English. Given the long history of these two resources and the economic challenges described below, the coverage of both the number of journal titles and of the various forms of literature vary to a significant degree, with the nadir being in the early 1980s when Beilstein, for example, indexed only 80 core organic chemistry journals. Beilstein started covering patents in 1869 and ceased that coverage in 1979.

Unlike most abstracting and indexing efforts or databases where the base record is a particular literature reference with attached indexing, both handbooks inverted this arrangement by making the substance the base record and then extracting information from all literature references for a given time period. The end result is that all information for a compound for a given time period appears in one place in the printed handbook. Entries are arranged by a detailed classification system. Such an arrangement naturally led to publication of extensive supplemental sets of volumes covering ever newer ranges of years. Beilstein was part way through its fifth supplement of the fourth edition covering literature from 1960–1979 before ceasing print volumes in 1998. Gmelin ceased publishing print volumes in 1997 when the entire institute was closed.

Since 1998, the history of Gmelin and Beilstein information is a tangled tale of acquisitions and intellectual property transfers. The Beilstein Institute no longer has any ownership stake in the electronic version of their data, with all rights ending up in the hands of Elsevier, the producer of the *Reaxys* system. Elsevier is also now the exclusive distributor of Gmelin content.

Entire books have been written about searching Beilstein, so a detailed discussion of the classification system and searching the print volumes is well beyond the scope of this chapter. However, some excellent brief guides to both resources have been created by science librarians for Beilstein[56,57] and Gmelin.[58,59] In addition, more extensive print books about Beilstein should still be available in major scientific libraries.[60,61] The key to searching the print handbooks is to identify the volumes the substance of interest appears in. In the case of Gmelin, which is published by chemical element, one must know the System Number of the element that takes priority in the compound. These system numbers appear inside the front cover of all but the oldest volumes. Volumes of Beilstein are published using a detailed hierarchical chemical structural classification system. For example, heterocyclics appear in volumes 17–27. Once one has the system/class number, one can go directly to the volume and page for the compound of interest since the same system number is used regardless of which time period (supplemental series) one is consulting.

6.5.4 Landolt Börnstein (Print)

Landolt-Börnstein (LB) is large set of critically reviewed data in physics, materials science, chemistry, astronomy, materials science, and related areas. The new electronic version of this resource is now branded as *SpringerMaterials* and is discussed in Section 6.4.2.4. The original German title is *Zahlenwerte und Funktionen aus Physik, Chemie, Astronomie, Geophysik und Technik*. The main sixth edition, published from 1950 to 1980 in 28 volumes, was continued by a large number of continuing supplemental volumes called the *New Series*. Solid state materials and semiconductors have received extensive coverage in the newer volumes.

Since many institutions have at least some of the print volumes but do not subscribe to *SpringerMaterials*, it is well worth the time to locate and search any print volumes that are available in one's own library or nearby large research libraries.

There is a comprehensive, single-volume print index for the sixth edition up through 1996. In addition, there is a free substance and property index[62] available on the Internet. Created by the University of Hamburg, the index covers 160,000 compounds. Molecular formula, CAS Registry Numbers, chemical names, substructure, and simple keyword search options are all available. The University of Texas Chemistry Library LB web guide[63] is one example of a well-crafted library guide providing more detailed information.

6.5.5 When to Consult the Primary Literature

As helpful as handbooks and other compilations of property data are, many physical properties are never extracted from the primary literature by these secondary sources. In addition, a property may indeed appear in a handbook, but is not discovered because the resource is obscure, not available, or simply missed by the searcher due the sheer number of handbooks published over the centuries.

After consulting some of the secondary sources described above and especially for less common compounds, it often is more efficient to try a search of the primary literature. Electronic databases and journal publisher web sites can often be used to zero in on a study reporting the property needed.

The advantage of traditional abstract and indexing databases such as *Chemical Abstracts or BIOSIS Previews*[64] is that an experienced indexer has added standard index terms and substance identifiers to the record, allowing the retrieval of a more detailed and targeted set of references than a search based merely on title and abstract keywords. The advantage of many journal publisher platforms is that they permit searching the full text of their articles, providing a level of detail that can find property information buried deep within the text. The disadvantage is that full text searching can often retrieve a very large number of references, many of them being irrelevant precisely because every word in the article is searched.

In either case, the trick is to design a search query that would retrieve only those references likely to measure and report the desired information. For example, vapor pressure or solubility data might well appear in an article discussing the environmental fate of a compound. A melting point or spectra may be given in a report of the synthesis of a chemical. Patents describing the use of a material in an electronic device may provide electrical data. Sometimes a search of the primary literature will quickly locate a reference because the exact data needed is the major focal point of the article and is clearly noted in the title, abstract, or indexing. At other times, one has to try a number of different queries and scan the full-text of a number of items before hopefully finding the required information. Chapters 2 and 3 of this book provide detailed information on searching the primary literature.

6.5.6 Conclusion

As noted in the introduction to this section, only a few representative examples of the thousands of high-quality print and electronic handbooks can be mentioned in this section.

It should be remembered that handbooks and other data compilations are fundamentally secondary sources. To be marketable, most tend to cover the more common properties of the more common substances. This can lead to the same data value being reported over and over again in dozens of handbooks, all from the same primary published research study. It is not uncommon to have situations where one wants the refractive index of a compound at 40 °C, only to find the same refractive index value at 20 °C. in many different handbooks. When faced with this situation, it often is a more effective strategy to directly search the primary literature using the large databases described in Section 6.4 or review in-depth library guides as described in the next section.

6.6 USING LIBRARY GUIDES AND SPECIALIZED SOURCES TO IDENTIFY ADDITIONAL RESOURCES

6.6.1 General Guides

Once the resources described in the previous four sections of this chapter have been exhausted or one is having trouble figuring out the right search process or specific resources, it is time to take advantage of guides prepared by chemical information professionals who, in many cases, work at large academic institutions.

Most are well-written guides that are carefully arranged and annotated, rather than a long list of miscellaneous links. Some also offer guidance in the search process itself, such as the *10 Easy Tips for Finding Property Data* (Arizona State University).[65] Another open source review is Chapter 13 of the *Chemical Information Sources* wikibook entitled "Physical Property Searches".[66] This is a successor to the classic Indiana University web site that was maintained for many years by Dr Gary Wiggins.

Poking around the web site of major science and engineering university libraries will often uncover excellent guides that contain much descriptive information on locating property and spectra data. A few good examples are listed below.

- *Thermodex: An Index of Selected Thermodynamic Data Handbook* (University of Texas Austin).[67]
- *Properties Data for Chemicals & Materials* (University at Buffalo).[68]

- *Finding Chemical & Physical Properties* (Vanderbilt University).[69]
- *Index to Physical, Chemical and Other Property Data* (Arizona State University).[70]
- *Chemical and Physical Properties* (Duke University).[71]

6.6.2 Finding Specialized Resources

Often the library guides described above will direct searchers to specialized resources that focus on a specific class of material or just a few types of properties or will be limited in both ways. For example, SpecialChem's *Additives Selector*[72] database is a resource covering a specific class while ONDA's *Tables of Acoustic Properties of Materials*[73] covers one type of property. Although most of the examples cited in this section are free web sites, print and subscription resources should not be neglected. Many chemists have had occasion to refer to classic sources such as *Azeotropic Data III*[74] based on Dow Chemical data and published by the American Chemical Society.

In addition, the power of general search engines such as Google in finding specialized property sources should not be underestimated. A search as simple as "thermal properties silicon" brings up at least seven good hits on the first page of its relevance ranked results. As with all web searches, one should not assume that the highest ranked results are necessarily accurate. One must always evaluate the quality of the information based on the organizational source, currency, and other documentation, such as whether references are given to the primary literature and the equipment and procedures are noted.

Some instructional guides on evaluating web sites indicate that ".edu" domain sites are automatically more trustworthy than ".com" sites. That is not necessarily the case with property and spectra sites. Some educational property sites are poorly designed and documented as well as having very limited content. Manufacturers and suppliers, mostly in the .com domain, usually have a vested interest in providing high-quality, accurate data to facilitate proper and safe use of their material. Evaluating data quality is more fully discussed in Section 6.3.

It would be impossible to make a list of all specialized resources. A few representative examples focusing on free web sites are provided below as an indication of the diversity of sources available.

(a) Focusing on specific classes of materials.
   (1) *ARS Pesticide Properties Database*.[75]
   (2) *Alternative and Advanced Fuel Properties Database*.[76]

    (3) *NIST Ceramics WebBook*[77]
    (4) *Plastics Technology Materials Database.*[78]
    (5) *Solv-DB.*[79]
    (6) *Aluminum Alloy Database*[80] (part of Knovel).
  (b) Focusing on specific properties.
    (1) *ATHAS Data Bank of Thermal Properties.*[81]
    (2) *Dielectric Constants of Various Materials.*[82]
    (3) *Handbook of Environmental Data on Organic Chemicals* (print[83] and electronic[84]).
    (4) *Thermophysical Properties of Matter*[85] – a massive 13-volume work covering thermal conductivity, specific heat, radiative properties, thermal diffusivity, viscosity, and thermal expansion.

Even these brief lists above emphasis the point that a comprehensive search requires accessing print, free web, and subscription resources.

6.6.3 Corporate Sources of Properties: Product Literature, Corporate Libraries and Archives, Notebooks

A wide variety of industries need and often develop extensive property information in the course of their normal business activities including basic and applied research, manufacturing, health and safety compliance, and marketing/sales efforts. Some of this is held for in-house use only, not necessarily because of competitive or confidentiality concerns, but rather a simple failure to publish the data in a publicly available form. Corporate employees usually have strong incentives to solve problems and fewer incentives to contribute data to scholarly publications. The exception may be data that will help the company sell its products.

Hence, product literature, including material safety data sheets (MSDS), technical specification sheets, and sales brochures in print or on the Internet, can be an invaluable source of property data. This is particularly true of trademarked products, mixtures, and other materials with variable or poorly defined compositions. An electrical conductivity value for a "pure" polymer tells you little about the conductivity of a given commercial compound with its assortment of additives. Sometimes the quickest approach to finding property information is to identify major manufacturers for the substance of interest using standard chemical directories. If a check of the sales literature at the company's web site is not productive, it is worth a call or email to the company's sales office. Marketing people are nearly always interested in assisting researchers and technologists that might discover new applications for their products which could lead to increased sales. At a minimum, the US material safety data sheet or its equivalent in other countries should be readily available. Of course, most MSDS contain a limited set of basic properties such as boiling point, vapor pressure, and specific gravity. However, MSDS can be a particularly good source of any property information related to safe handling, such as flash point, solubility, and reactivity.

For those working for corporations that research or manufacture materials, the best place to start the search is by getting to know your company's information specialists working in the library, archives, records department, and marketing department. Do not assume that all information staff work in easily identifiable departments such as the "research library". They may be attached to business intelligence or product testing units. A searcher should also recognize that the data they need may have never been extracted into an easily accessible compilation, like a company data book or major technical report. The required data may be buried in a hard copy or an electronic notebook, a monthly progress report, or old, outdated product literature. Sales departments tend to keep only current product literature on hand. Corporate information centers or archives are often the only sources for older data. This is all the more reason to develop a strong relationship with information professionals within your organization that can help you access what can be termed the "corporate memory".

6.7 FINDING SPECTRA

Spectroscopy is a broadly used analytical approach to identifying materials by recording the absorption or emission of electromagnetic radiation across the entire spectrum. Mass spectrometry, which detects the masses of fragmented compounds, is also included. It is important to have access to well-documented sets of reference spectra for comparison in identifying unknown substances, confirming reaction products, and characterizing components in mixtures or functional groups of compounds. As with other resources, they can be divided into two camps: fee-based and free.

Spectra are also widely reported in the primary literature as part of experimental results. Hence, sometimes the fastest way to find spectra is to do a search of primary literature using a large database such as *SciFinder*. Spectral analysis software packages are also available to manage, analyze, and transform experimental spectra and compare them against reference libraries, but these are beyond the scope of this chapter. Laboratory personnel know that modern instruments often come with large libraries of reference spectra and many software tools, which are an obvious place to start one's search, when available.

As has been noted throughout this chapter, many of the major resources cover both physical properties and spectra. Rather than repeat the detailed descriptions of these major resources, they will be listed below alphabetically, with a brief mention of the types of spectra covered and cross-referenced to the relevant chapter section.

- *ChemSpider*[18] – Spectra from the open web can be found under the main heading *Spectra* and on the *Spectral Data* tab under the *Data Sources* main heading (Section 6.4.1.2).
- *Knovel*[41] – Not only is there spectral data embedded as figures and tables in many of the handbooks, but a few titles are filled with spectra, *e.g.*, *Rapra Collection of Infrared Spectra of Rubbers, Plastics and Thermoplastic Elastomers, 3rd Edition*[86] (Section 6.4.2.3).
- *NIST Chemistry WebBook*[25] – IR spectra, mass spectra, UV/Visible spectra, and vibrational and electronic spectra with searching by compound and energy levels. Though the number of spectra is not especially large, as with all NIST data, it is critically evaluated (Section 6.4.1.1).
- *Reaxys*[39] – a particularly rich source for spectral data, though often one must retrieve the full text of the original literature reference to see the actual graphical spectra (Section 6.4.2.2).
- *SciFinder/Chemical Abstracts Databases*[30,32,34] – The REGISTRY substance record may have an extensive list of experimental spectra, some of which is digitized, and calculated (predicted) ^{13}C and proton NMR spectra for most organics. Failing that, one must search the primary literature in the *CAplus* file which automatically includes MEDLINE,[38] if using the *SciFinder* platform (Section 6.4.2.1).

The remaining subsections will cover databases devoted to spectral information.

6.7.1 Best and Largest Free Spectral Databases

6.7.1.1 Spectra Data Base System for Organic Compounds.[87] The *Spectra Data Base System for Organic Compounds* (SDBS) is organized and maintained by the National Institute of Advanced Industrial Science and Technology (AIST) of Japan. SDBS includes electron impact mass, Fourier transform infrared, proton NMR, ^{13}C NMR, laser Raman, and electron spin resonance spectra in varying availability for some 34,000 organic compounds, mostly measured by AIST. A number of the collections are still being updated. The system can be searched by compound name, molecular formula, molecular weight, CAS Registry Number, atom count, and spectra peaks or shifts. The site is free to use, but requires acknowledging an agreement to abide by their requested terms of use.

6.7.1.2 Sigma-Aldrich Product Catalog Advanced Search.[88] The *Sigma-Aldrich* online catalog has freely available links to NMR, IR, FTIR, and Raman spectra for many of the compounds they sell. However, those links are somewhat hard to find. One must do a search, click on a product number, and then click on the *Safety and Documentation* tab. If there are links to spectra, they will appear in the left-hand *Documents* column. Unfortunately, the only documentation for the spectra seems to be for the solvent, if any, and the instrument used. The full name of the instrument is not given. For example, one NMR spectra examined simply had "QE300" above the spectra as shorthand for a General Electric QE-300 spectrometer. The spectra are not dated and are assumed to have been mostly produced in-house.

6.7.1.3 FTIRsearch.com/RAMANsearch.com.[89] Under this brand name, Thermo Scientific[90] offers a pay-per-search, curated database of over 71,000 FTIR and nearly 16,000 Raman spectra spanning many classes of compounds. One can submit an unknown spectra for pattern matching against the database. A text search option searches by chemical name, CAS Registry Number, molecular formula, and other text fields. The product is designed as an affordable option for smaller organizations and occasional searchers.

6.7.2 Best and Largest Subscription Spectral Databases

6.7.2.1 KnowItAll (Bio-Rad).[91] In 1978, Bio-Rad acquired Sadtler Research Laboratories, the publisher of very large print collections of spectra (See Section 6.7.3.1), renaming it as their Informatics Division and launching a full line of electronic products. Spectral databases, which can be licensed individually, include general and specialized collections in ATR, IR, MS, NMR of several elements, Raman, and UV/Visible. The special collections cover technical areas such as flame retardants and lubricant additives.

Of particular interest are their two comprehensive bundles, *KnowItAll Enterprise* for corporate users and *KnowItAll U* for academic users. Those interested in determining their exact needs, license terms, and cost should contact the company. *KnowItAll* includes over 1.4 million spectra from a variety of sources, including Bio-Rad, Sadtler, Wiley, and SDBS (Section 6.7.1.1). Both the enterprise and academic version permit building internal, centralized databases of spectra. *KnowItAll AnyWare* is the web-based interface used to access the databases and includes compound searching, peak searching, pattern matching, and analysis. Finally, the *KnowItAll Informatics System* can also be licensed and downloaded for more advanced analysis, structure drawing, and reporting. A free teaching version, branded *KnowItAll Academic Edition* and available to individual academic users, includes the *KnowItAll Informatics System* along with IR and NMR spectra for 1,300 compounds.

6.7.2.2 SpecInfo (Wiley).[92] Wiley spectra, in addition to being licensed to KnowItAll and to Chemical Abstracts (Section 6.4.2.1), are available through Wiley's own *SpecInfo* system that uses the *SpecSurf* search and analysis software to process extensive sets of NMR, IR, and MS spectra. Users can search by or input structures, spectra or keywords, analyze spectra against the reference collections, and predict spectra for structures drawn by the user.

6.7.3   Print Resources, Library Guides, and Specialized Resources

The number of print and electronic resources focusing on particular types of spectra or classes of compounds is rather overwhelming. There are far too many to itemize here, which is an indication of how important spectroscopy is to chemical research. Many of the classic print resources are now quite dated. Caution should be exercised in using spectra run on what are now obsolete, lower-resolution instruments. Not all sources provide as extensive data provenance for their spectra as would be desirable. Of course, they may still have value in the identification of unknowns.

Now that many modern instruments come with built-in spectral libraries of tens of thousands of compounds and with open data sources developing on the Internet, the need to consult published sources has diminished. Those serving as chemical information specialists/librarians can testify that the number of requests they receive for spectra has decreased greatly over the past 20 years.

6.7.3.1 Print Sources. Despite their age, two very large historic print compilations of spectral data must be mentioned since they may well contain older unusual compounds, trademarked material, and other commercial substances no longer readily available, if at all. These volumes might well be residing on print reference shelves or in remote storage facilities of larger research libraries. Again, a proper search of one's library catalog should readily identify what portions of these sets are owned by one's institution. Specific spectra may also be cited by number in older scholarly publications.

Sadtler Spectra: Almost certainly the most extensive collection of print spectra ever created was produced by Sadtler Research Laboratories, the predecessor to Bio-Rad's extensive electronic product line (Section 6.7.2.1). Those long familiar with chemistry libraries will likely remember several shelves of these in dark green, 3-ring binders in the reference section. Sadtler collections were also produced on microfilm. Most of what they published fell into two categories: large standard spectra collections and highly focused commercial collections such as agricultural chemicals or plasticizers. Standard collections ran the gamut of grating and prism IR, FTIR, ^{1}H NMR, ^{13}C NMR, DTA, ATR, IR, and UV spectra. Usually commercial collections covered the more common IR or ^{1}H NMR spectra, although there are DTA and ATR spectra for polymers.

Thermodynamic Research Center (TRC) Selected Spectral Data: The TRC has a long history starting with the National Institute of Science and Technology (NIST) in 1942, then many years at Texas A&M University before rejoining NIST in 2000.[93] From about 1945 through 2000, they produced many spectra collections, including major projects for the American Petroleum Institute and the Manufacturing Chemists Association, now the American Chemistry Council. IR, UV, MS, ^{1}H and ^{13}C NMR, and Raman spectra were all issued in dozens of loose-leaf binders. As with Sadtler, much of the spectra is dated but retains some value. Unlike Sadtler, these spectra have not been reissued in electronic form. However, TRC has reissued much of its thermodynamic property data as NIST databases (Section 6.4.1.1).

6.7.3.2 Library Guides and Specialized Resources. The best approach to finding specialized resources is to consult library guides at major research institutions, previously described in Section 6.6 for property information. A search of one's library catalog is also highly recommended. A small sample of high-quality library guides is given below:

- Chemistry: Spectra & Spectral Data (University of Arizona).[94]
- Spectra & Spectral Data (University at Buffalo)[95] – two tabs: by spectra type and by compound class.
- Spectral Information Resources (Stanford University).[96]
- Spectroscopy (University of Wisconsin-Madison).[97]

Finally, a very selective and merely representative list of specialized resources is provided below to give searchers a small sense of the diversity of material available:

- **Atomic and Microwave Spectra:** *Diatomic Spectral Database* (NIST).[98]
- **Infrared:** *Handbook of infrared and Raman spectra of inorganic compounds and organic salts* (Academic Press)[99] – Published in four print volumes.
- **Mass Spectra:** *NIST/EPA/NIH Mass Spectral Library with Search Program*[100] – Over 240,000 MS and significant GC retention data. Available only from distributors as a stand-alone Windows PC software package.
- **NMR:** *NMRShiftDB*[101] – Freely available content with registered users contributing content. Originally developed by the collaborative efforts of individuals at a number of European institutions.
- **UV/Visible:** *Handbook of Ultraviolet and Visible Absorption Spectra of Organic Compounds* (Knovel)[102] and *Absorption Spectra in the Ultraviolet and Visible Region* (24 vols., Academic Press).[103]
- **Geosamples:** *USGS Digital Spectral Library* (U.S. Geological Survey)[104] – UV/Near-IR spectra of minerals, mixtures, artificial, liquids, volatiles, and vegetation.
- **Packaging Additives**: *Spectra for the Identification of Additives in Food Packaging* (Kluwer Academic)[105] – FT-IR, MS, and ^{1}H NMR spectra.

The mass spectra library and UV/visible handbook mentioned in above listing are also available in print.

6.7.4 Crystallography

X-ray crystallography, long used to study inorganic compounds, has developed into a key tool for looking at the crystalline structures of semiconductors, organic compounds, and biomolecules. Most famously, the elucidation of the structure of DNA was based on an x-ray diffraction pattern produced by Rosalind Franklin[106] in 1952 and interpreted correctly by Watson and Crick.

Of the sources previously discussed, the CAS databases (Section 6.4.2.1), SpringerMaterials/Landolt Börnstein (Sections 6.4.2.4 and 6.5.4), and Reaxys/Gmelin/Beilstein (Sections 6.4.2.2 and 6.5.3) have particularly significant amounts of crystallographic data. For example, in the detailed *SciFinder* substance display, various properties such as crystal lattice parameters, pore sizes, and XRD patterns are listed in the *Structure-related Properties* section of the *Experimental Data*. In addition, when retrieving literature references for any substance or group of substances, one can limit the search to the "Crystal Structure" role, which was assigned from 1967 forward. However, the search capabilities of these more general resources are generally not as sophisticated in searching crystallographic information as are the databases dedicated to x-ray data, nor do they provide data analysis software tools.

Traditionally, the collection and curation of x-ray crystallography has been the domain of scientific societies and centers. Most of these organizations also provide a variety of data analysis and visualization computer programs to enhance the value of the underlying data.

6.7.4.1 International Tables for Crystallography (ITC). Since its founding in 1948, the International Union of Crystallography (IUCr) in Chester, England has published a series of reference volumes entitled the *International Tables for Crystallography*.[107] These volumes contain critically evaluated tables and reviews on all aspects of crystallographic research, including symmetry, techniques, and properties of crystals in general. To date, eight volumes (A-G and A1), some in multiple editions, have been published. An electronic version[108] is also available for purchase from the Union. A helpful guide, *Symmetry in Crystallography: Understanding the International Tables*, was published in 2011.[109]

6.7.4.2 Cambridge Structural Database (CSD/WebCSD). The Cambridge Crystallographic Data Centre (CCDC) in Cambridge, England, compiles and distributes the *Cambridge Structural Database* (CSD),[110] a major international repository of experimentally determined organic and metal-organic crystal structures. The database contains 3D structures of over 500,000 x-ray and neutron diffraction analyses, either from the published literature or directly deposited by researchers. Over 40,000 structures are added each year, each carefully validated by independent reviewers. Various software packages enhance the value of CSD. Both in-house and web versions of the database are available by subscription.

CCDC provides a free teaching database of 500 structures.[111] The 3D nature of the structures and the geometry of their interactions with other molecules and ions can be visualized in a number of different display modes. Teaching exercises show how the interactive database can be used to teach concepts such as aromaticity, VSEPR, and stereochemistry.

6.7.4.3   Inorganic Crystal Structure Database (ICSD).   Produced by
FIZ Karlsruhe in Eggenstein-Leopoldshafen, Germany, the *ICSD*[112]
contains about 150,000 peer-reviewed entries with atomic coordinates
back to 1913. Included are crystal structures for elements and compounds of up to five elements. About 7,000 new structures are added
each year, and existing records are undergoing continuing review. In
partnership with the National Institute of Standards and Technology
(USA), legacy data on metallic and intermetallic compounds is being
added. ICSD can be accessed by subscription via the Web, CD-ROM,
or intranet options.

6.7.4.4   Powder Diffraction File (PDF).   The International Centre for
Diffraction Data (ICDD) in Newtown Square, Pennsylvania (USA) are
the producers of the Powder Diffraction File.[113] Until 1978, ICDD was
known as the Joint Committee on Powder Diffraction Standards
(JCPDS). The 2011 release of the Powder Diffraction File has almost
750,000 data sets containing diffraction, crystallographic, and bibliographic information. Records contain experimental details and selected
physical properties. All data is evaluated and annotated as to quality.
Minerals, organics, modulated structures, polymers, and clays are covered, provided they have a reproducible diffraction patterns. Various
subsets of the data can be subscribed to as needed.
   The licensing/access model is complex and limited, due in part to the
sophisticated suite of computer programs that are required to access the
data sets. Although there are site licenses that go beyond the typical
single user/single computer mode, a site is limited to a maximum of 10
users at a single building. Access from more than a single computer
requires the user to use a USB software protection dongle. There are no
site-wide licenses permitting any computer in an institution access to the
database.

6.7.4.5   Open Data Sources.   Recently, some researchers and organizations have wanted to make crystallographic data more generally
available and have developed open data initiatives. In the interest of
space, some of the more notable efforts are listed below along with
brief annotations. In most cases, the annotations are taken directly
from the referenced web sites. Sites providing data analysis programs
and utilities are beyond the scope of this chapter. As noted in the
introduction to the chapter, resources dealing solely with biomolecules
are not included.

   • *American Mineralogist Crystal Structure Database*[114] – Provides
     access to every structure published in *American Mineralogist, The*

Canadian Mineralogist, the European Journal of Mineralogy, and the Physics and Chemistry of Minerals, as well as selected datasets from other journals. The database is maintained by the Mineralogical Society of America and the Mineralogical Association of Canada with funding from the National Science Foundation (USA).

- *Crystallography Open Database*[115] – A public domain, open access collection of over 120,000 crystal structures of organic, inorganic, metal-organic compounds, and minerals, excluding biopolymers. The web site provides little documentation about this resource and its history. However, it apparently is an effort spearheaded by two academic institutions in Vilnius, Lithuania, and was funded primarily by the Research Council of Lithuania.

- *Database of Zeolite Structures*[116] – This database provides structural information on all of the zeolite framework types that have been approved by the Structure Commission of the International Zeolite Association. It includes descriptions and drawings of each framework type, user-controlled animated displays, crystallographic data, simulated powder diffraction patterns for representative materials, and bibliographic references.

- *Reciprocal Net*[117] – Still under development, this distributed database is a cooperative effort of 18 crystallography laboratories, mostly in North America. The emphasis is on structures of general interest to teachers, students, and the general public, with most of the data being open access. In line with its educational purposes, the project is funded by the National Science Foundation (NSF) and is part of the National Science Digital Library.[118]

6.7.4.6 Other Sources. For information on additional subscription and open data resources, consult:

- The *Data Activities in Crystallography* web page maintained by the International Union of Crystallography.[119]
- *Crystallographic Databases*, a detailed classroom handout prepared by Dana Roth of the California Institute of Technology Millikan Library that reviews all the major databases.[120]

Much of the older data in the *Cambridge Structural Database* and the *Inorganic Crystal Structure Database* is also available in the serial, *Strukturbericht* (1913–1939) and its successor title, *Structure Reports* (1940–1993). This newer title was split in two in 1965, with Section A covering metals and inorganic compounds and Section B covering organic and organometallic compounds.

6.8 CONCLUSION

Beyond becoming familiar with the resources described in this chapter, important skills for a searcher are persistence, creativity, and a sense of what avenues are most likely to be successful and which ones are unproductive. This is not unlike the qualities of a good detective. Indeed, searchers could well be described as "information detectives". Good searchers must be willing to explore different strategies within and across resources. They should also be willing to ask for help from more experienced searchers. Not all searches end in success. However, thoughtful selection and careful searching of the sources described in this chapter should certainly increase the chances of finding the exact data one needs.

REFERENCES

1. E. W. Washburn, C. J. West and C. Hull, *International Critical Tables of Numerical Data, Physics, Chemistry and Technology*, 1st edn., Pub. for the National Research Council by the McGraw-Hill Book Co., New York, 1926.
2. http://www.proquest.com/en-US/catalogs/databases/detail/pqdt.shtml. Accessed 6 June 2012.
3. http://worldwide.espacenet.com/. Accessed 6 June 2012.
4. R. S. Boethling, P. H. Howard and W. M. Meylan, *Environ. Toxicol. Chem.*, 2004, **23**, 2290–2308.
5. C. Hochstein, S. Arnesen and J. Goshorn, *Medical Reference Services Quarterly*, 2007, **26**, 21–45.
6. http://sis.nlm.nih.gov/enviro/toxweblinks.html. Accessed 6 June 2012.
7. P. J. Andree, M. F. Harper, S. Nauche, R. A. Poolman, J. Shaw, J. C. Swinkels and S. Wycherley, *World Pat. Inf.*, 2008, **30**, 300–308.
8. D. Edwards, J. E. Stajich and D. Hansen, *Bioinformatics Tools and Applications*, New York : Springer, 2009.
9. E. Go, *Journal of Neuroimmune Pharmacology*, 2010, **5**, 18–30.
10. T. Klingstrom and D. Plewczynski, *Briefings Bioinf.*, 2011, **12**, 702–713.
11. R. A. Laskowski, *Mol. Biotechnol.*, 2011, **48**, 183–198.
12. T. O'Grady, *College & Research Libraries News*, 2008, **69**, 404–421.

13. W. J. Geldenhuys, K. E. Gaasch, M. Watson, D. D. Allen and C. J. Van der Schyf, *Drug Discovery Today*, 2006, **11**, 127–132.
14. http://tigger.uic.edu/~mansoori/Thermodynamic.Data.and.Property_html. Accessed 8 August 2013.
15. http://ccd.chemnetbase.com. Accessed 27 July 2012.
16. http://www.cas.org/expertise/cascontent/registry/regsys.html. Accessed 6 June 2012.
17. http://www.iupac.org/home/publications/e-resources/inchi.html. Accessed 6 June 2012.
18. http://www.chemspider.com/. Accessed 6 June 2012.
19. http://commonchemistry.org/. Accessed 6 June 2012.
20. *Compendium of Chemical Terminology (Gold Book Version 2.3.1)*, Research Triangle Park, NC International Union of Pure and Applied Chemistry, 2012. http://goldbook.iupac.org/. Accessed 6 June 2012.
21. I. Mills, *Quantities, Units, and Symbols in Physical Chemistry*, 2nd edn., Boston, Oxford, 1993. http://old.iupac.org/publications/books/gbook/green_book_2ed.pdf. Accessed 6 June 2012.
22. http://itlaw.wikia.com/wiki/Data_provenance. Accessed 8 June 2012.
23. http://www.ceramics.nist.gov/IDELA/IDELA.htm. Accessed 7 June 2012.
24. R. G. Munro, *Data Evaluation Theory and Practice for Materials Properties*, U.S. Dept. of Commerce, Technology Administration, National Institute of Standards and Technology, [Gaithersburg, Md.], 2003. http://purl.fdlp.gov/GPO/gpo3836. Accessed 8 June 2012.
25. http://webbook.nist.gov/chemistry/. Accessed 6 June 2012.
26. http://srdata.nist.gov/gateway/. Accessed 6 June 2012.
27. http://trc.nist.gov/. Accessed 15 June 2012.
28. http://www.matweb.com/index.aspx. Accessed 6 June 2012.
29. http://www.chemexper.com/. Accessed 6 June 2012.
30. http://www.stn-international.de/uploads/tx_ptgsarelatedfiles/CAPLUS_02.pdf. Accessed 11 June 2012.
31. http://www.dialog.com/. Accessed 11 June 2012.
32. http://www.cas.org/products/scifindr. Accessed 8 August 2013.
33. http://www.stn-international.org/. Accessed 11 June 2012.
34. http://www.cas.org/File%20Library/Training/STN/DBSS/registry.pdf. Accessed 8 February 2013.
35. http://www.spresi.com/. Accessed 11 June 2012.

36. http://www.cas.org/File%20Library/Training/STN/User%20Docs/propsources.pdf. Accessed 8 February 2013.
37. http://www.acdlabs.com/home/. Accessed 11 June 2012.
38. http://www.cas.org/File%20Library/Training/STN/DBSS/medline.pdf. Accessed 8 February 2013.
39. https://www.reaxys.com/info/. Accessed 6 June 2012.
40. http://www.elsevier.com/online-tools/reaxys/training-and-support. Accessed 8 August 2013.
41. http://why.knovel.com/. Accessed 06 June 2012.
42. E. W. Washburn and Knovel Corp., *International Critical Tables of Numerical Data, Physics, Chemistry and Technology*, 1st Electronic edn., 1926, 2003. http://www.knovel.com/web/portal/browse/display?_EXT_KNOVEL_DISPLAY_bookid=735. Accessed 26 July 2012.
43. Knovel Corp., *Knovel Critical Tables*, 2nd edn., 2008. http://www.knovel.com/web/portal/browse/display?_EXT_KNOVEL_DISPLAY_bookid=761. Accessed 8 August 2013.
44. http://www.springermaterials.com/navigation/. Accessed 6 June 2012.
45. http://library.caltech.edu/learning/classhandouts/Landolt-Bornstein_SpringerMaterials-2011.pdf. Accessed 25 July 2012.
46. http://www.aiche.org/dippr/events-products. Accessed 8 February 2013.
47. http://i-systems.dechema.de/detherm/. Accessed 6 June 2012.
48. https://cindasdata.com/. Accessed 6 June 2012.
49. https://cindasdata.com/products. Accessed 06 June 2012.
50. R. E. Maizell, *How to Find Chemical Information : A Guide for Practicing Chemists, Educators and Students*, 4th edn., Wiley-Blackwell, Oxford, 2009.
51. M. G. Mellon, *Chemical Publications, their Nature and Use*, 5th edn., McGraw-Hill, New York, 1982.
52. http://www.hbcpnetbase.com/. Accessed 6 June 2012.
53. http://www.crcnetbase.com/. Accessed 6 June 2012.
54. http://products.asminternational.org/hbk/index.jsp. Accessed 6 June 2012.
55. G. W. C. L. T. H. Kaye, *Tables of Physical and Chemical Constants*, 16th edn., New York, Essex, England, 1995. http://www.kayelaby.npl.co.uk/. Accessed 6 June 2012.
56. http://www.lib.lsu.edu/sci/chem/guides/srs147.html. Accessed 6 June 2012.
57. http://www.lib.utexas.edu/chem/info/beilstein.html. Accessed 6 June 2012.
58. http://www.lib.utexas.edu/chem/info/gmelin.html. Accessed 6 June 2012.

59. http://library.buffalo.edu/libraries/asl/guides/gmelin.html. Accessed 6 June 2012.

60. Beilstein Institute, *How to Use Beilstein : Beilstein Handbook of Organic Chemistry*, Frankfurt/Main, Beilstein Institute, 1978.

61. O. Weissbach and F. K. Beilstein, *The Beilstein Guide : A Manual for the Use of Beilsteins Handbuch der Organischen Chemie*, Springer-Verlag, Berlin; New York, 1976.

62. http://lb.chemie.uni-hamburg.de/. Accessed 6 June 2012.

63. http://www.lib.utexas.edu/chem/info/lb.html. Accessed 6 June 2012.

64. http://thomsonreuters.com/biosis-previews/. Accessed 8 August 2013.

65. http://www.asu.edu/lib/noble/chem/prophelp.htm. Accessed 6 June 2012.

66. http://en.wikibooks.org/wiki/Chemical_Information_Sources/Physical_Property_Searches. Accessed 6 June 2012.

67. http://www.lib.utexas.edu/thermodex/. Accessed 6 June 2012.

68. http://library.buffalo.edu/libraries/asl/guides/engineering/properties.html. Accessed 6 June 2012.

69. http://researchguides.library.vanderbilt.edu/Property. Accessed 8 August 2013.

70. http://www.asu.edu/lib/noble/chem/property.htm. Accessed 6 June 2012.

71. http://library.duke.edu/research/subject/guides/chemical-physical-properties/index.html. Accessed 6 June 2012.

72. http://www.specialchem4polymers.com/product-directory/index.aspx. Accessed 6 June 2012.

73. http://www.ondacorp.com/tecref_acoustictable.shtml. Accessed 6 June 2012.

74. L. H. Horsley and W. S. Tamplin, *Azeotropic Data III*, American Chemical Society, Washington, D.C., 1973.

75. http://www.ars.usda.gov/services/docs.htm?docid=14199. Accessed 7 June 2012.

76. http://www.afdc.energy.gov/afdc/fuels/properties.html. Accessed 7 June 2012.

77. http://www.ceramics.nist.gov/webbook/evaluate.htm. Accessed 7 June 2012.

78. https://www.ptonline.com/plaspec/. Accessed 7 June 2012.

79. http://solvdb.ncms.org/solvdb.htm. Accessed 7 June 2012.

80. http://www.knovel.com/web/portal/browse/display?_EXT_KNOVEL_DISPLAY_bookid=844. Accessed 7 June 2012.

81. http://athas.prz.rzeszow.pl/Default.aspx?op=db. Accessed 7 June 2012.

82. http://www.deltacnt.com/99-00032.htm. Accessed 8 February 2013.

83. K. Verschueren, *Handbook of Environmental Data on Organic Chemicals*, 5th edn., John Wiley & Sons, Hoboken, N.J., 2009.

84. http://www.knovel.com/web/portal/browse/display?_EXT_ KNOVEL_DISPLAY_bookid=2437. Accessed 8 August 2013.

85. Y. S. Touloukian, *Thermophysical Properties of Matter, IFI/* Plenum, New York, 1970–1979.

86. M. Forrest, Y. Davies and J. Davies, *Rapra Collection of Infrared Spectra of Rubbers, Plastics and Thermoplastic Elastomers (3rd Edition)*, Smithers Rapra Technology. http://www.knovel.com/web/ portal/browse/display?_EXT_KNOVEL_DISPLAY_bookid= 3087. Accessed 8 August 2013.

87. http://riodb01.ibase.aist.go.jp/sdbs/cgi-bin/cre_index.cgi?-lang=eng. Accessed 8 August 2013.

88. http://www.sigmaaldrich.com/catalog/AdvancedSearchPage.do. Accessed 11 June 2012.

89. https://ftirsearch.com. Accessed 8 August 2013.

90. http://www.thermoscientific.com/. Accessed 8 August 2013.

91. http://www.knowitall.com/literature/. Accessed 12 June 2012.

92. http://onlinelibrary.wiley.com/book/10.1002/9780471692294. Accessed 12 June 2012.

93. http://trc.nist.gov/history.html. Accessed 15 June 2012.

94. http://lro.library.arizona.edu/subject-guide/252-Chemistry? tab=10079. Accessed 8 August 2013.

95. http://library.buffalo.edu/libraries/asl/guides/spectra.html. Accessed 11 June 2012.

96. http://www-sul.stanford.edu/depts/swain/help/subjectguides/spectral. html. Accessed 11 June 2012.

97. http://chemistry.library.wisc.edu/subject-guides/spectroscopy.html. Accessed 8 February 2013.

98. http://physics.nist.gov/PhysRefData/MolSpec/Diatomic/index.html Accessed 11 June 2012.

99. R. A. Nyquist, C. L. Putzig, M. A. Leugers, R. O. Kagel and R. A. Nyquist, *The Handbook of Infrared and Raman Spectra of Inorganic Compounds and Organic Salts*, Academic Press, San Diego, 1997.

100. http://www.nist.gov/srd/nist1a.cfm. Accessed 11 June 2012.

101. http://www.nmrshiftdb.org; http://nmrshiftdb.nmr.uni-koeln.de/. Accessed 8 August 2013.

102. K. Hirayama, *Handbook of Ultraviolet and Visible Absorption Spectra of Organic Compounds, Plenum Press/*Springer, New York, 1967.

103. L. Lang, *Absorption Spectra in the Ultraviolet and Visible Region*, Academic Press, New York, 1961.

104. http://speclab.cr.usgs.gov/spectral-lib.html. Accessed 11 June 2012.

105. B. v. Lierop, *Spectra for the Identification of Additives in Food Packaging*, Kluwer Academic, Boston, 1998.

106. S. Rapoport, *The History Teacher*, 2002, **36**, 116–127.

107. http://www.mx.iucr.org/iucr-top/it/. Accessed 10 June 2012.

108. http://it.iucr.org/. Accessed 10 June 2012.

109. P. G. Radaelli, *Symmetry in Crystallography: Understanding the International Tables [IUCr Texts on Crystallography v. 17]*, New York, Oxford, 2011.

110. http://beta-www.ccdc.cam.ac.uk/Solutions/CSDSystem/Pages/CSD.aspx. Accessed 10 June 2012.

111. http://www.ccdc.cam.ac.uk/Solutions/FreeSoftware/Pages/CSDTeachingDatabase.aspx. Accessed 8 August 2013.

112. http://www.fiz-karlsruhe.de/icsd.html. Accessed 10 June 2012.

113. http://www.icdd.com/products/. Accessed 10 June 2012.

114. http://www.geo.arizona.edu/AMS/amcsd.php. Accessed 10 June 2012.

115. http://www.crystallography.net/. Accessed 10 June 2012.

116. http://www.iza-structure.org/databases/. Accessed 10 June 2012.

117. http://www.reciprocalnet.org/. Accessed 10 June 2012.

118. http://nsdl.org/. Accessed 16 June 2012.

119. http://www.iucr.org/resources/data. Accessed 10 June 2012.

120. http://library.caltech.edu/learning/classhandouts/Crystallographic Databases.pdf. Accessed 12 June 2012.

CHAPTER 7

Commercial Availability, Safety, and Hazards

DANA L. ROTH

California Institute of Technology, Millikan 1-32, Pasadena, CA 91125, US
Email: dzrlib@library.caltech.edu

7.1 INTRODUCTION

The importance of being able to obtain chemicals commercially, being aware of necessary laboratory precautions, and having access to hazardous-property information is essential for students and researchers. This chapter provides commercial availability links to individual chemical company catalogs and to websites that are linked to multiple commercial sources. It describes laboratory-safety information resources and provides hazardous-property information, for both individual and classes of chemicals, and includes appropriate search techniques.

Information on commercial availability, safety, and hazards is interrelated in the sense that commercial chemical companies are responsible for providing Material Safety Data Sheets (MSDSs) for their products. The MSDS for a substance provides basic physical-property information, possible hazardous properties, and instructions for safe use.

7.2 COMMERCIAL AVAILABILITY

Individual supplier catalogs, such as Aldrich Chemistry, as well as many catalogs including the products of multiple suppliers, such as Chemical

Chemical Information for Chemists: A Primer
Edited by Judith N. Currano and Dana L. Roth
© The Royal Society of Chemistry 2014
Published by the Royal Society of Chemistry, www.rsc.org

Book, are freely available online. In addition, both SciFinder and Reaxys provide subscribers with links from individual compound records to several suppliers or supplier catalogs. Given the enormous number of supplier websites for common chemicals (*e.g.*, search Google for "suppliers of benzene"), serious consideration should be given to finalizing searches using well-established, up-to-date, individual supplier catalogs, since some suppliers do not actually stock chemicals but claim to make them on demand.

In general, use of CAS Registry Numbers is recommended, since they are unique, rather than chemical names or molecular formulas, which can be ambiguous. It should be noted that each stereochemical or isotopic variant of a compound has a unique CAS Registry Number (CAS RN).

7.2.1 Commercial Availability Resources

7.2.1.1 Individual Company Sites. **Aldrich Chemistry**[1] offers a full range of chemistry products for a variety of research areas, including asymmetric syntheses, catalysis, chemical biology, greener alternatives, organometallics, solvents, stable isotopes, *etc.* This website also includes products from the Fluka and Sigma lines. The catalog is both text and structure searchable. The default text search is limited to either product names/numbers or CAS Registry Numbers. Structure searching offers the option of three structure editors (JME, Marvin, ChemDraw), plus the ability to limit by ranges of MW, BP, or MP. The advanced search offers a wide variety of "search type" and "search method" options for text, plus the ability to perform keyword searches of the Sigma-Aldrich web pages, PDFs, DataSheets, technical bulletins, brochures, *etc.*

The product records provide a link for downloading the Material Safety Data Sheet for the substance (which presumably updates the 1988 Sigma-Aldrich library of chemical safety data), and generally include brief physical property data, safety information, references to the Merck Index, Beilstein, Fieser & Fieser, and the Aldrich spectra catalogs. They also provide a links out to PubChem Substance and may link to PDF images of IR or NMR spectra. In addition to a listing of research grades, there may also be a link to bulk ordering and pricing options.

7.2.1.2 Multiple Supplier Sites. **Chemical Book**[2] is a freely available site for commercial chemicals, with options for an English, Chinese, German, or Japanese interface. The default search is "Products", which is searchable with a CAS Registry Number or a chemical name/ synonym. Since companies are not always careful about assigning the current CAS RN, it may be necessary to perform a second search using a chemical name (for example, Pyrogallol Red) to ensure complete results.

The "Chemical site" option provides short descriptions of the nearly 50 companies, most of which are located in China. Chemical products are also browsable by category (*e.g.*, Amino Acid Derivatives, Catalysis & Inorganic Chemistry, *etc.*). It is also possible to browse CAS Registry Numbers, the chemical name index or the company website index (which lists thousands of international companies and gives links to a seemingly random listing of their chemical products).

Search results include a structural diagram, basic physical properties, and links to both China and global suppliers, chemical properties (basic information, physical properties, safety information, usage and synthesis, preparation products and raw materials, *etc.*) and MSDS providers.

ChemNet[3] is a freely searchable listing of chemicals and related products from over 26,000 worldwide suppliers. The default home page "Product Search" box allows searching by product name, CAS RN, or molecular formula. Data for individual compounds, however, do not seem as extensive as Chemical Book. Additional search options are Company Search (keyword searchable), ChemSearch (keyword web search), and Directory Search (keyword site search).

In addition to the ChemNet home page, there are tabs for pages that allow one to browse or search through products, suppliers, offers to buy and sell, chemical news, REACH, and CAS RNs. The "Resource" tab has a list of links to chemical categories, such as colour index, drug synthesis, FEMA (Flavor chemicals) index, Open Access journals, pesticide common names, and FDA databases.

MatWeb[4a] provides searchable data sheets for over 90,000 polymers, plastics, metals, alloys, ceramics, semiconductors, fibers, and engineering materials. Each data sheet provides an extensive listing that includes a material description, physical, mechanical, thermal, and processing properties. MatWeb allows searchers to identify suppliers for both specific materials and discovered materials with specific properties.

Free registration is recommended in general,[4b] but it is required to access the Advanced Search Engine, which offers iterative searching. Search options include text, property data (physical, mechanical, *etc.*), material category (carbon, ceramic, metal, *etc.*). The advanced search option is extremely powerful; for example, one can use it to perform a search for substances that are alloys, have tensile stresses within a specific range of values, and contain a nonferrous metal.

MatWeb also includes several other search tabs. "Category" allows you to browse and search by material categories. "Property" offers a combination search of material category and material properties, which includes links to "Polymer Film Search" and "Lubricant Search". "Metals" presents a combination search of material category and material composition. The remaining tabs, "Trade Name" and "Manufacturer," are self-explanatory.

The ChemExper Chemical Directory[5] is a freely available catalog consisting of more than 9 million chemicals from more than 2,300 suppliers. The default "Quick Search" accepts molecular formula, CAS Registry Numbers, product names or synonyms, and SMILES strings. Structure searching, both exact and substructure, is available with the JME Molecular Editor. The "Advanced Search" option offers a combination of IUPAC name (equals, starts with, contains), molecular formula, and structure, as well as inequality and range relations for numeric fields, such as boiling point, melting point, density, and IR absorption frequencies. Availability of NMR spectra can also be specified. Records for specific chemicals include the following information: CAS RN, molecular formula, molecular weight, boiling point, density, cLogP, cLogS, polar surface, 3D model viewer, InChIKey, *etc.*

SciFinder[6] (subscription required) has links, in REGISTRY, to CHEMCATS, a database that indicates commercial availability of substances. At the time of this writing, CHEMCATS has more than 67 million products (based on ~ 20 million unique chemical substances) found in $\sim 1,000$ supplier catalogs. While some records do include quantities and pricing, others do not, and it is always wise to contact the supplier directly for the most up-to-date pricing, availability, quantities, purities, *etc.*

Reaxys[7] (subscription required) has links, indicated by a flask icon under the structural diagram, to Accelry's ACD (license required), eMolecules (freely available), and Cambridge Soft ChemACX (subscription required). eMolecules[8] can also be freely searched directly using either chemical structures or names. Pricing and use of the database for screening purposes requires a subscription.

7.3 SAFETY

7.3.1 The Importance of Safety in the Chemical Laboratory

The American Chemical Society issued "The Chemical Professional's Code of Conduct"[9] in 2007. It lists a number of safety-related issues, specifically, that chemical professionals should "actively be concerned with the health and safety of co-workers, consumers and the community," that "they should strive to understand and anticipate the environmental consequences of their work," and that "they have a responsibility to minimize pollution and to protect the environment".[9]

Unfortunately, readers of *Chemical & Engineering News* are reminded, on an almost weekly basis, of violations of the "Code of Conduct" in both the "Chemical Safety" Letters to the Editor[10], and in Jyllian Kemsley's "The Safety Zone" blog,[11] which provides a "Friday Chemical Safety Round-up," including sections on chemical health and safety news from the past two weeks, fires and explosions, leaks, spills, and other exposures. Most of the time, however, it does not specifically include stories related to meth labs; ammonia leaks; incidents involving floor sealants, cleaning solutions, or pool chemicals; transportation spills; and fires from oil, natural gas, or other fuels. In addition to the Safety Round-up, the "Safety Zone" has several other sections, including Popular Posts, Recent Posts, Recent Comments, Lessons Learned Reports, Blogroll, and Chemical Safety Resources.

This dichotomy, between the "Code" and the "Safety Zone" reports, strongly suggests that everyone working with chemicals should increase their awareness of possible hazards in their work, employing both print and electronic resources to learn more about both the dangers of working with chemicals and protective measures that should be taken. The pervasive nature of the general problem of safety, especially in academia, was recently recognized by the ACS Committee on Chemical Safety (CCS), which recently convened a task force that produced a draft report, "Creating Safety Cultures in Academic Institutions,"[12] which is available on the CCS website. This draft report is focused on identification of elements of strong safety cultures, safety education, and enhancement of safety cultures.

7.3.2 Safety Resources

In addition to developing a general culture of safe handling of chemicals, it is also essential to maintain an on-going awareness of statutory regulations related to their use. Two essential publications help meet these needs; Material Safety Data Sheets,[13] available from chemical suppliers,[14] provide safe-handling procedures, health hazards, exposure limits, and basic physical properties, while the very practical publication "Prudent Practices in the Laboratory: Handling and Disposal of Chemicals"[15] has chapters on "The Culture of Laboratory Safety", "Working with Chemicals", "Government Regulation of Laboratories", and other, general topics related to laboratory safety.

In addition to freely available electronic resources, there are also a variety of print resources, easily available in many university and large public libraries, including:

R. H. Hill, Jr., D. C. Finster, *Laboratory safety for chemistry students*, Wiley, 2010.

A. K. Furr, *CRC handbook of laboratory safety, 5th edition*, CRC Press, 2000.

On being a scientist: A guide to responsible conduct in research. National Academies Press, 2009. http://www.nap.edu/catalog.php?record_id=12192, Accessed May 24, 2013.

R. Montesano ... W. Davis, *Handling chemical carcinogens in the laboratory:* Problems *of safety,* IARC Scientific Publications; no. 33, Who Publications Center, 1979.

Safety in academic chemistry laboratories, 7th ed., American Chemical Society, Joint Board-Council Committee on Chemical Safety, 2003.

Volume 1: Accident prevention for college and university students, http://portal.acs.org/portal/PublicWebSite/about/governance/committees/chemicalsafety/publications/WPCP_012294, Accessed May 24, 2013.

Volume 2: Accident prevention for faculty and administrators, http://www.chem.uoa.gr/misc/SACL_faculty.pdf, Accessed May 24, 2013.

N. V. Steere, *Safety in the chemical laboratory*, v.1–4. Reprinted from *J. Chem.* Educ., American Chemical Society Division of Chemical Education, 1967–1981.

It can be challenging to effectively search for the most up-to-date information on hazards and safety; the *Chemical Information Sources*[16] wiki book provides additional tips on performing such searches for safety information, and any librarian or information professional will also be able to provide assistance in this area.

7.4 HAZARDS

7.4.1 Introduction

The Chemical Abstracts Service currently adds approximately 15,000 new substances to its Registry File each day,[17] and controversies over the assessment of chemical risks in the United States continue. While only relatively few of these substances will enter the environment, current federal regulations are widely recognized as inadequate, especially for nanoparticles and hormone mimics,[18] two classes of compounds that are especially insidious since, while they may not be toxic, they can have deleterious effects on pulmonary function and fetal development.

The Toxic Substances Control Act,[19] which became law in 1976, regulates the introduction of new or existing chemical substances in the market place. It is flawed, however, in the sense that it did not provide a mechanism for dealing with the possible toxicity of substances already on the commercial market. The Act mandates that the EPA maintain the TSCA Chemical Substance Inventory,[20] which is a listing of existing chemicals. Chemicals not listed generally must be reviewed by the EPA prior to their commercial use.

This review process, however, is questionable and apparently subject to external pressures. A Chicago Tribune[21] investigation of flame retardant chemicals suggests that these pose significant health risks and are generally ineffective. It also identifies highly questionable legislative testimony regarding the necessity of adding toxic chemicals to products, especially those destined for infants.

Recent positive developments include the EPA's expansion of the use of computational toxicology in regulatory decisions about commercial chemicals and pesticides, as well as unregulated contaminants in drinking water. This effort, called "ToxCast", focuses on how chemicals cause health effects and promises to reduce the use of animal testing.[22]

7.4.1.1 Nano Particles. Nano particles are of special concern because their hazard potential was originally based on the material's bulk properties. This unfortunate situation is only recently being recognized, and steps are rapidly being taken to fully explore the difference in the chemical nature of, for example, gold nanoparticles as contrasted with metallic gold.[23]

The remarkable diversity of new nanomaterials is causing serious consideration of reinventing toxicological testing.[24] Currently, there is an enormous variety of potential nanomaterials, but no standards for determining which are possibly toxic to humans and other species. This situation is an obvious threat to public confidence in new products and, if continued, will certainly restrict their development. Standard toxicological approaches, which require testing of individual materials, are thought to be an inadequate response to the demands in the current economic climate; so, new approaches are being tested.[25] In 2012, the current state of the art of both the science and the significant data gaps posed by engineered nanomaterials was described in a National

Research Council publication entitled, "A research strategy for environmental, health, and safety aspects of engineered nanomaterials."[26]

An additional problem that arises when attempting to retrieve toxicity information about nanoparticles is that they have only recently been identified as such in the Chemical Abstracts Service's CAplus records. The CAS Registry does not separately register either nanomaterials or nanoscale objects (as defined by the external dimensions of the material). These materials are indexed with the same CAS RN as the bulk material. This leads to a complicated search strategy; for example, searching *Chemical Abstracts* for the toxicity of a beryllium nanoparticle on Sci-Finder requires a search for the element (*e.g.*, Substance Identifier=Beryllium; CAS RN=7440-41-7, *etc.*), retrieving the references limited by the substance role (Adverse Effect, including toxicity), and then refining by research topic with the word "nano".

In SciFinder/STN, a NANO super role has been added for those substances, which have been indexed in records since 1992, if they are described as being either:

- in a nano form or as nanoscale without further information; or
- between 0.05 and 100 nanometers in one dimension.

Tox Town, the National Library of Medicine's interactive guide to commonly encountered toxic substances, is a resource that provides general information, and has released a Nanoparticles page.[27]

7.4.1.2 Hormone Mimics. Hormone mimics are another serious problem. A fairly detailed account of the effects of estrogenic mimics is given in the Wikipedia article on Endocrine disruptors.[28] A recent study,[29] funded by the National Institute of Environmental Health Sciences, quantified the biological effects of estrogenic activity, and found that about 95 percent of commercial plastic items tested positive for estrogenic activity, although not all are toxic. As pointed out in a recent review,

"Exposure in people is typically a result of contamination of the food chain, inhalation of contaminated house dust or occupational exposure. EDCs include pesticides and herbicides (such as dichlorodiphenyl trichloroethane or its metabolites), methoxychlor, biocides, heat stabilizers and chemical catalysts (such as tributyltin), plastic contaminants (*e.g.*, bisphenol A), pharmaceuticals (*i.e.*, diethylstilbestrol; 17α-ethinylestradiol) or dietary components (such as phytoestrogens)".[30]

Sources of information for hazardous materials are widely varied. Academics provide the basic research which is reported in journal articles and comprehensively indexed in PubMed,[31] industrial organizations that have concerns about the health of their employees and any legal obligations governing both industrial use and commercial products require access to a variety of database resources, and a wide variety of government agencies have statutory responsibilities for both research and enforcement of regulations, as reflected in the Federal Register,[32] which is excerpted in ChemList (subscription required).[33]

7.4.2 Hazards Resources

Searching for information on hazardous chemicals is generally very straightforward, due to the universal use of Chemical Abstracts Registry Numbers for compound identification. Fortunately, there are a wide variety of both subscription and freely available databases, which update and continue many print resources that remain useful for the compounds they describe.

7.4.2.1 Hazards Resources – Subscription Required. **Bretherick's Handbook of Reactive Chemical Hazards**[34] covers some 5,000 elements and compounds. Information on the instability of individual compounds and hazardous interactions between elements and/or compounds appears in volume one; classes, groups of elements, and compounds possessing similar structures or hazard potential are alphabetically listed in volume two. Entries are in chemical formula order, with extensive cross-indexing. Source material is based on both journal articles and book series on synthetic techniques. These sources are supplemented by encyclopedia works such as Mellor, Houben-Weyl, and Kirk-Othmer, as well as *Chemical Abstracts,* and the two RSC publications, *Laboratory Hazards Bulletin* and *Chemical Hazards in Industry.* Print, Kindle, and other online versions of the handbook are available, and Bretherick's forms the foundation of Elsevier's HazMat Navigator database. The following appendices are included.

- Source title abbreviations used in references.
- Tabulated fire-related data.
- Glossary of abbreviations and technical terms.
- Index of chemical names and synonyms used in section 1.
- Index of class, group and topic titles used in section 2.
- Index of section 2 titles classified by type.
- Index of CAS registry numbers *vs.* serial numbers in section 1.

Hazmat Navigator[35] is a chemical safety database based on Bretherick's Handbook of Reactive Chemical Hazards, with additional content from Sittig's Handbook of Toxic and Hazardous Chemicals and Carcinogens, Sittig's Handbook of Pesticides and Agricultural Chemicals, Encyclopedia of Toxicology, and the Fire Protection Guide to Hazardous Materials. In addition to critical, detailed chemical hazards information from Bretherick's, Hazmat Navigator includes information on physical properties, handling, safety, flammability and toxicity from both the journal literature and other resources. This information is updated annually.

The search page offers the following search options:

- Quick: chemical name or synonym, CAS RN or molecular formula.
- Advanced: keyword database search or fielded search.
- Structure/substructure: draw structures using the MarvinSketch structure editor.
- Index: alphabetical listing of contents with links.

The element/compound record provides links to:

- Properties: names, properties, RTECS #, Code #s, emolecules, *etc.*
- Handling: recommendations for PPE, spill, storage, disposal, *etc.*
- Safety: common uses, first aid, exposure guidelines, *etc.*
- Reactivity: common reactive hazards, links to related records.
- Flammability: fire and explosion hazards, firefighting procedure.
- Toxicity: human acute toxicity, target organs, environmental fate, *etc.*

A demonstration video with scrollable sample pages from the Benzene record is available at http://tinyurl.com/ych68rk.

An interesting and very important feature of Hazmat Navigator is the link to related records. For example, the record for Benzvalene (a very strained Benzene isomer), has only partial information (*i.e.*, a brief listing of properties and a short reactive hazard note), but it also has a related record link to Strained-Ring Compounds, a listing of similar compounds and compound groups with Hazmat records.

Sax's Dangerous Properties of Industrial Materials[36] currently lists more than 26,000 substances, an increase of approximately 10% over the previous edition. In addition, approximately 1,000 chemicals have been updated to include IDLH (Immediately Dangerous Life or Health) levels. Sax is unique in the number of chemicals listed and the breadth of information on toxic and mutagenic potential, flammable and explosive properties, reactivity and physical properties, and regulatory data. Print indexes include synonym lists and CAS Registry Numbers. Sax is available in print, on CD-ROM, on the Wiley Online Library, and in Knovel. Wiley Online search options include full text, article title, CAS RN, and chemical name

Patty's Industrial Hygiene and Toxicology Online[37] is the electronic version of the sixth edition of *Patty's Industrial Hygiene* and the Sixth edition of *Patty's Toxicology*. *Patty's Industrial Hygiene* is focused on the engineering control of exposure to chemical, physical, and biological hazards in the workplace. *Patty's Toxicology* provides toxicological information for chemical compounds used in industry, including metals, organics, and polymers. Patty's online is available in the Wiley Online Library, searchable as combined or separate volumes, and in Knovel. Wiley Online search options include full text, article title, CAS RN or chemical name, and topic.

***Dictionary of Substances and their Effects* (DOSE)**[38] covers over 4,100 chemicals, including pesticides, food carcinogens, and endocrine disruptors. Information is provided on physical properties, exposure limits, and toxicity in general, as well as geno- and ecotoxicity, environmental fate, and regulatory requirements. The second edition appeared in seven volumes in 1999, and purchase of the print volumes includes site-wide access to DOSE Online, a fully searchable web database. Knovel, however, provides access to the third edition (2005), which covers ~5,300 compounds. Search options include chemical name/synonym, CAS RN, and EINECS/RTECS number. The RSC website, http://www.rsc.org/Publishing/CurrentAwareness/DOSE/SampleRecords.asp, provides a sample record.

7.4.2.1.1   Chemical Hazard Information Library. The Chemical Hazard Information Library (CHIL)[39] is an online collection of full text documents from over 100 national and international sources, including: EPA (Environmental Protection Agency), NLM (National Library of Medicine), EU (European Union), WHO (World Health Organization, NTP (National Toxicology Program), ATSDR (Agency for Toxic Substances and Disease Registry), NCI (National Cancer Institute), IARC (International Agency for Research on Cancer), DoD (Department of Defense), *etc.* It contains toxicology and related hazard data affecting both humans and the environment. The complexity of this database and the extent of its source materials require a greatly expanded discussion of search options, as well as examples of the use of these options.

1. **New Search/Basic Search** is the default search screen, with links to Search Options (which offers "Apply related words" and "... search full text of the articles" and various limits), Advanced Search (which is Basic Search plus Search Options), Visual Search (which presents results with subject headings for limiting), and Search History. Phrase searching is the default, along with the Boolean search options: AND; OR; and NOT. Fielded searching (*e.g.,* author, molecular formula, name of substance, source collection, title words, *etc.*) is also available. Click "Advanced Search", which shows a "Field Codes" link just above the search box for a listing of Field Code tags.

   Example 1a: A New Search (Basic Search) for 'Dioxin' retrieves over 17,000 results.

   Example 1b: Entering Dioxin as a keyword and then clicking Advanced Search offers the additional options of including "Apply related words" and "Also search within the full text of the articles", which now retrieves over 21,000 results. Adding a Field Code (TI dioxin) now retrieves the 12,000 results that have "dioxin" as a title word. Results can be refined by faceted options in the left margin: Full Text; PDF Full Text; Subject; and Publication. There is also a "date" scale for additional limiting.

   Example 1c: A search for "Dioxin" in Visual Search provides an interesting "Group Results" list of 250 results with a choice of either refining with subject terms or providing a list of publications.

2. **EXPERTIndex** defaults to a searchable/browsable listing of more than 2.25 million names & synonyms. Clicking on a compound name/synonym generates a search of the CHIL compound ID number, within Search History, and provides options to View Results (a listing of documents), View Details, or Edit (which allows access to Search Options).

   Example 2: A browse/search for "Dioxin" in EXPERTindex retrieves 4920 results. The default is both "related words" and "full text", and the results can by refined by faceted options in the left margin. The EXPERTINdex search searches the UI field for the UI code. These documents are "about" dioxin, as opposed to being documents containing the word "dioxin".

3. **Collections** is a browsable listing of all the collections in CHIL. Clicking on a collection name generates a search which allows adding additional search terms and limits results to that collection.

Example 3: A Collections search for "CPDB – Carcinogenic Potency Database AND dioxin" (where dioxin is searched as a text word) retrieves three results, which, in this case, provide extensive experimental detail on carcinogenicity.

4. **ChemID Enhanced** is a searchable/browsable listing of names and synonyms. Clicking on a compound name displays a structure diagram, basic physical properties, synonyms, classification codes, RTECS descriptors, and Chemical and Physical Properties from the Hazardous Substances Data Bank. Clicking on the name of a substance in the ChemID Enhanced record generates an EXPERTIndex search.

Example 4: A search for "Dioxin" in ChemID Enhanced retrieves a record with the chemical structure, molecular formula, CAS RN, Synonyms, etc., and HSDB chemical and physical properties. Clicking on the chemical name link generates an EXPERTindex search, that can be modified with additional criteria, such as "plastics", to limit retrieval to three results.

5. **More**. Clicking on Indexes offers the option to browse various indexes (*e.g.*, Category, Collection, Document Type, Published Date) and add search terms.

7.4.2.2 Hazards Resources – Freely Available. **TOXNET**. The TOXNET[40] fact sheet states that it "is a group of databases covering chemicals and drugs, diseases and the environment, environmental health, occupational safety and health, poisoning, risk assessment and regulations, and toxicology" and that information in the TOXNET databases covers:

- specific chemicals, mixtures, and products;
- chemical nomenclature;
- unknown chemicals;
- special toxic effects of chemicals in humans and/or animals; and
- citations from the scientific literature.

The primary TOXNET databases of interest to chemists are ChemIDplus, HSDB, Toxline, and the Household Products Database.

ChemIDplus[41] is a web-based search system that serves as the chemical index for TOXNET. ChemIDplus records provide structure descriptors (InChI, Smiles), chemical synonyms, toxicity values, physical properties and Locator Code links to a wide variety of other government agency sponsored files (NLM, FDA, EPA, Superlist Locator, *etc.*) and databases (PubMed, PubChem, *etc.*).

At the time of this writing, the database contains over 388,000 chemical records, of which over 295,000 include chemical structure diagrams, and is searchable by Name, Synonym, CAS Registry Number, Molecular Formula, Classification Code, Locator Code, Structure, Toxicity, and/or Physical properties.

The ChemIDplus Lite version is available for simplified Name and CAS RN searching without the need for plugins or applets. ChemIDplus Advanced allows searching for chemical structures and substructures (MarvinSketch), toxicity, and physical properties. An enhanced structure display is also available by clicking the Enlarge Structure link.

The Hazardous Substances Data Bank (HSDB)[42] is the most comprehensive online resource for hazardous substances. It is freely available and provides peer-reviewed toxicological information for over 5,400 potentially hazardous chemicals. HSDB records for individual chemicals provide data in the following categories.

- Human Health Effects.
- Emergency Medical Treatment.
- Animal Toxicity Studies.
- Metabolism/Pharmacokinetics.
- Pharmacology.
- Environmental Fate/Exposure.
- Environmental Standards & Regulations.
- Chemical/Physical Properties.
- Chemical Safety & Handling.
- Occupational Exposure Standards.
- Manufacturing/Use Information.
- Laboratory Methods.
- Special References.
- Synonyms and Identifiers.
- Administrative Information.

A unique feature of the HSDB are the extensive references to a core set of books, government documents, technical reports, and the primary journal literature.

HSDB search terms include chemical names and name fragments, synonyms, CAS RN. The text is keyword searchable (*e.g.*, chromium compounds, kidney failure, oil spill, *etc.*). Searching for a specific chemical (*e.g.*, Benzene) displays a link to the primary Benzene record as well as links to other records that mention Benzene. Clicking the primary Benzene link displays the "Human Health Effects" section, with the full table of contents in the left margin. If you click on one of the other records or perform a keyword searching for a topic, the default display will be the Best Section (*i.e.*, section with the highest term frequency).

TOXLINE[43] is a bibliographic database covering the biochemical, pharmacological, physiological, and toxicological effects of drugs and other chemicals. It currently has over 4 million records with abstracts, indexing terms and CAS Registry Numbers. TOXLINE sources include PubMed, reports and archival material from other government agencies, and other TOXNET databases. Search results offer: relevancy ranking; sorting; and downloading options.

The Household Products Database[44] provides information on potential health effects of chemicals found in more than 10,000 products used inside and around the home. Users can browse product categories (*e.g.*, Pesticides), sub-categories (*e.g.*, Insecticide), or types (*e.g.*, ants), which then display an alphabetical listing of product names (*e.g.*, Ant Stop Orthene Fire Ant Killer); or, one can scan alphabetical lists of product names (*e.g.*, Alumaseal), types of products (*e.g.*, Antifreeze), manufacturers (Airwick Industries), and ingredients (*e.g.*, Acetone).

There are also both a quick search and an advanced search option for product name, type, manufacturer, ingredient (searchable by chemical name or CAS Registry Number), and health effects, all of which can also be limited by product category.

Clicking on the Product Name for a record displays product information, including a customer service phone number and links to "products with similar usage", and manufacturer information, including "Related Items: Products by this Manufacturer" and health effects, handling and disposal, and ingredients information from the manufacturer's Material Safety Data Sheet (MSDS), as well as a link to the complete MSDS. It is also possible to highlight terms and initiate a search in the TOXNET database.

The International Programme on Chemical Safety (IPCS),[45] in partnership with CCOHS, was developed to establish a scientific rationale for management of chemicals, and to ensure chemical safety of both humans and the environment. The IPCS publication program (http://www.inchem.org/) covers both natural and manufactured chemicals and exposure situations ranging from chemicals in the natural environment to their extraction or synthesis, industrial production, transport, use, and disposal.

The Nanomaterial-Biological Interactions Knowledgebase (NBIK)[46] is designed to provide a scientific basis for understanding the effects of nanomaterial exposure in biological systems. The NBIK collects information on characteristics of nanomaterials (purity, size, shape, charge, composition, functionalization, agglomeration state); methods of synthesis; and beneficial, benign or deleterious nanomaterial-biological interactions at the organism, cellular, and molecular level. The NBIK database will allow for accurate definition of nanomaterial structure-activity relationships, and these representations can be used to predict nanomaterial properties in the absence of empirical data.

The KNOWLEDGEBASE link provides access to both a Nanomaterial Library and a Biological Interactions Database, which have drop-down menus for searching by material type, shape, core, charge, surface Chemistry, size (min/max), and dendrimer generation. Additional information links are provided for the following.

- Goals and Objectives of the NBI.
- Biological Implications of Nanotechnology.
- Organizational Chart.
- Frequently Asked Questions.

ACToR (Aggregated Computational Toxicology Resource)[47] is a publicly available database focused on high and medium production volume industrial chemicals, pesticides (both active and inert ingredients), and potential ground and drinking water contaminants. It provides access to over 1,000 resources and information on chemical toxicity and potential chemical risks for over 500,000 chemicals. ACToR is searchable by chemical names (keyword or exact), CAS Registry Numbers, and chemical structures searched as both substructures and similarity, *via* a ChemAxon structure editor.

Search results (after clicking the "details" link) include basic physical property data, synonyms, toxicology data (hazard, acute toxicity, chronic toxicity, carcinogenicity, genetic toxicity, ecotoxicity), category data (risk management, exposure data), and a list of external databases searched by name or CAS RN.

ChemSpider[48] is a freely available database, based on chemical structures, that provides information on over 26 million de-duplicated compounds derived from over 400 sources. These sources include a wide variety of government databases, chemical supplier catalogs, and academic and commercial websites. Each of the listed sources[49] has a brief pop-up description, with the full record providing a web link to the source. ChemSpider augments the default information provided by these sources with a very wide variety of additional of data, including:

- Properties/Experimental: Presents flammability, toxicity, and safety data.
- RSC Databases: Click on "Laboratory Hazards Bulletin" for journal article titles, author names, and article DOIs.
- Medical Subject Headings Classification: MeSH heading and PubMed terms and heading to which the substance maps.

Bear in mind that: "Before you do anything with ChemSpider it is important to remember that it isn't yet either perfect or exhaustive."[50] ChemSpider search options include:

1. **Simple:** Systematic name; synonym; trade name; Registry Number; SMILES, InChI, or CSID, with additional options for single/ multi-component, isotopically labeled, and additional filters.
2. **Structure:** Upload a structure or image file; convert name, SMILES, InChI, or ChemSpider ID to structure; or draw a structure (Accelrys JDraw, Elemental, ACD/Labs SDA, Ketcher, JME, or JChempaint). Searches can be exact, substructure, or similarity, with additional options for exact match, all tautomers, same skeleton (including or excluding H), and all isomers.
3. **Advanced:** Search by structure, identifier, elements, properties, calculated properties, data source, *etc.*, and by LASSO (Ligand Activity in Surface Similarity Order) similarity.
4. **Search History:** Each search is assigned an internal ID number, date/time started and updated, status, progress, predicate (search string), and system message.

7.4.3   Conclusion

This section provides an overview of significant resources for hazardous chemicals and chemical products. Given the growing importance of this problem in the United States, new publications and resources will surely continue to appear on a regular basis. Thieme Medical Publishers has prepared a comprehensive listing of hazard resources for "the different hazardous properties that Copy Editors should take into consideration when annotating experimental procedures",[51] which also includes both US, UK and European legislation related to hazards and laboratory safety.[51]

Sources for continuing education are weekly reading of *Chemical & Engineering News* and maintaining awareness of new publications and databases (*e.g., Exposure Science in the 21st Century: A Vision and a Strategy,*[52] ChemHat.org,[53] and NOAA's *Chemical Reactivity Worksheet.*[54]

REFERENCES

1. Aldrich Chemistry, http://www.sigmaaldrich.com/chemistry.html, Accessed April 16, 2013.
2. Chemical Book, http://www.chemicalbook.com/ProductIndex_EN.aspx, Accessed April 16, 2013.
3. ChemNet : Global Chemical Network, http://www.chemnet.com/, Accessed April 16, 2013.
4. (a) MatWeb ... Source for Materials Information, http://www.matweb.com/, Accessed April 16, 2013; (b) Benefits of registering with MatWeb, http://www.matweb.com/membership/benefits.aspx, Accessed April 16, 2013.
5. ChemExper Chemical Directory, http://www.chemexper.com/, Accessed April 16, 2013.
6. Chemical Suppliers - CHEMCATS - Find commercially available chemicals, pricing, and supplier contact information, http://www.cas.org/expertise/cascontent/chemcats.html, Accessed April 16, 2013.
7. Reaxys, https://www.reaxys.com/info/, Accessed April 16, 2013.
8. eMolecules, http://www.emolecules.com/, Accessed April 16, 2013.
9. American Chemical Society. The Chemical Professional's Code of Conduct, http://portal.acs.org/portal/PublicWebSite/careers/profdev/ethics/CNBP_023290, Accessed May 24, 2013.
10. Chemical Safety: Nitric Oxide at High Pressure, *Chemical & Engineering News*, 2012, 90, 6, http://cen.acs.org/articles/90/i5/Chemical-Safety-Nitric-Oxide-High.html, Accessed May 24, 2013.
11. Central Science: The Safety Zone, http://cenblog.org/the-safety-zone/, Accessed April 17, 2013.
12. Creating Safety Cultures in Academic Institutions, www.acs.org/safety, Accessed April 16, 2013.
13. Material safety data sheet. *Wikipedia, the Free Encyclopedia*, http://en.wikipedia.org/wiki/Material_safety_data_sheet, Accessed April 17, 2013.
14. Links to materials safety data sheets are available in commercial chemical supplier catalogs, some of which are described in Section 7.2.
15. National Research Council of the National Academies, *Prudent practices in the laboratory: Handling and management of chemical*

hazards, National Academies Press, Washington, D. C., 2011. http://www.nap.edu/catalog.php?record_id=12654, Accessed May 24, 2013.

16. Chemical safety and toxicology searches. In *Chemical Information Sources*, http://en.wikibooks.org/wiki/Chemical_Information_Sources/Chemical_Safety_Searches, Accessed April 17, 2013.

17. CAS Registry and CAS Registry Number FAQs. CAS (Chemical Abstracts Service), http://www.cas.org/content/chemical-substances/faqs, Accessed May 25, 2013.

18. P. Hunt. Letters: Assessing Chemical Risk: Societies Offer Expertise, S*cience*, 2011, 131, 1136, http://www.sciencemag.org/content/331/6021/1136.1.full.pdf, Accessed May 24, 2013.

19. Toxic Substances Control Act of 1976, *Wikipedia, the Free Encyclopedia*, http://en.wikipedia.org/wiki/Toxic_Substances_Control_Act_of_1976, Accessed May 22, 2013.

20. TSCA Chemical Substance Inventory, http://www.epa.gov/oppt/existingchemicals/pubs/tscainventory/index.html, Accessed May 22, 2013.

21. P. Callahan, S. Roe. Big Tobacco wins fire marshals as allies in flame retardant push, *Chicago Tribune* Watchdog, May 8, 2012, http://www.chicagotribune.com/news/watchdog/flames/ct-met-flames-tobacco-20120508,0,3332088.story, Accessed May 24, 2013.

22. C. Hogue. *New Tools for Risk Assessment, Chemical & Engineering News,* 2012, 90, 32, http://cen.acs.org/articles/90/i24/New-Tools-Risk-Assessment.html, Accessed May 24, 2013.

23. T. Masciangioli, J. Alper. *Challenges in characterizing small particles: exploring particles from the nano- to microscales:A workshop summary*, National Academies Press, 2012, http://www.nap.edu/catalog.php?record_id=13317, Accessed May 24, 2013.

24. R. F. Service. Can high-speed tests sort out which nanomaterials are safe? *Science*, 2008, 321,1036, doi: 10.1126/science.321.5892.1036, Accessed May 24, 2013.

25. S. George, T. Xia, R. Rallo, Y. Zhao, Z. Ji, S. Lin, X. Wang, H. Zhang, B. France, D. Schoenfeld, R. Damoiseaux, R. Liu, S. Lin, K. A. Bradley, Y. Cohen and A. E. Nel, *ACS Nano*, 2011, 5, 1805, doi: 10.1021/nn102734s, Accessed May 24, 2013.

26. National Research Council. *A research strategy for environmental, health, and safety aspects of engineered nanomaterials*. National Academies Press, 2012, http://www.nap.edu/catalog.php?record_id=13347, Accessed May 24, 2013.

27. ToxTown – Nanoparticles, http://www.toxtown.nlm.nih.gov/text_version/chemicals.php?id=67, Accessed May 22, 2013.

28. Endocrine disruptor, *Wikipedia, the Free Encyclopedia*, http://en. wikipedia.org/wiki/Endocrine_disruptor, Accessed May 22, 2013.

29. C. Z. Yang, S. I. Ganiger, V. C. Jordan, D. J. Klein, G. D. Bittner. *Environ Health Perspect*, 2011, 119, 989, http://ehp03.niehs.nih.gov/ article/info:doi%2F10.1289%2Fehp.1003220, Accessed May 24, 2013.

30. C. A. Frye, E. Bo, G. Calamandrei, L. Calzà, F. Dessi-Fulgheri, M. Fernández, L. Fusani, O. Kah, M. Kajta, Y. Le Page, H. B. Patisaul, A. Venerosi, A. K. Wojtowicz and G. C. Panzica. *J. Neuroendocrinol*, 2012, 24, 144, doi: 10.1111/j.1365-2826.2011. 02229.x., Accessed May 24, 2013.

31. PubMed, http://www.ncbi.nlm.nih.gov/sites/entrez?otool=caitlib, Accessed May 22, 2013.

32. *Federal Register* – The Daily Journal of the U.S. Government, https://www.federalregister.gov/, Accessed May 22, 2013.

33. Regulated Chemicals – CHEMLIST, http://www.cas.org/content/ regulated-chemicals, Accessed May 22, 2013.

34. P. Urben. *Bretherick's Handbook of Reactive Chemical Hazards, 7th ed.* Amsterdam; London: Elsevier, 2006. Work consists of two volumes: v.1: Introduction; Reactive chemical hazards; Specific Chemicals (in formula order), and v.2: Classes, Groups and Topics (in alphabetical order).

35. N. Langeman, R. J. Alaimo, F. Fung. *Hazmat Navigator*, http:// www.elsevierdirect.com/hazmatnavigator/home.html, Accessed May 25, 2013. *Hazmat Navigator* is a chemical safety database based on Bretherick's Handbook of Reactive Chemical Hazards (subscription required/free trial available).

36. R. J. Lewis. *Sax's Dangerous Properties of Industrial Materials, 12th ed.*, New York: John Wiley & Sons, 2013. http://onlinelibrary. wiley.com/book/10.1002/0471701343, Accessed May 25, 2013.

37. Patty's Industrial Hygiene and Toxicology Online, http:// onlinelibrary.wiley.com/book/10.1002/0471125474, Accessed May 25, 2013. Includes *Patty's Industrial Hygiene, 6th Ed.*, edited by Vernon E. Rose (2011) and *Patty's Toxicology, 6th ed.*, edited by Eula Bingham (2012).

38. S. Gangolli. *Dictionary of Substances and their Effects (DOSE), 2nd ed.* Cambridge: Royal Society of Chemistry, 1999. Available online thru Dialog (309), Knovel, and the RSC.; About DOSE, http:// www.rsc.org/Publishing/CurrentAwareness/DOSE/About.asp, Accessed May 25, 2013.

39. CHIL: Chemical Hazard Information Library (subscription required/free trial available), EBSCO Discovery Service, http://

www.ebscohost.com/biomedical-libraries/chemical-hazard-information-library, Accessed May 25, 2013.

40. (a) TOXNET: TOXicology Data NETwork, National Library of Medicine Fact Sheet: http://www.nlm.nih.gov/pubs/factsheets/toxnetfs.html Database links: http://toxnet.nlm.nih.gov/, Accessed May 25, 2013. (b) TOXNET and Beyond: Using the National Library of Medicine's Environmental Health and Toxicology Portal, National Library of Medicine, 2011, http://sis.nlm.nih.gov/enviro/toxnetmanualfeb2011.pdf, Accessed May 25, 2013.

41. ChemIDplus, National Library of Medicine Factsheet: http://www.nlm.nih.gov/pubs/factsheets/chemidplusfs.html, Accessed May 25, 2013. Database (Lite): http://chem.sis.nlm.nih.gov/chemidplus/chemidlite.jsp, Accessed May 25, 2013. Database (Advanced): http://chem.sis.nlm.nih.gov/chemidplus/, Accessed May 25, 2013.

42. Hazardous Substances Data Bank (HSDB), National Library of Medicine Fact Sheet: http://www.nlm.nih.gov/pubs/factsheets/hsdbfs.html, Accessed May 25, 2013. Data Bank: http://toxnet.nlm.nih.gov/cgi-bin/sis/htmlgen?HSDB, Accessed May 25, 2013.

43. TOXLINE, National Library of Medicine Factsheet: http://www.nlm.nih.gov/pubs/factsheets/toxlinfs.html, Accessed May 25, 2013. Database: http://toxnet.nlm.nih.gov/cgi-bin/sis/htmlgen?TOXLINE, Accessed May 25, 2013.

44. Household Products Database, National Library of MedicineFactsheet: http://www.nlm.nih.gov/pubs/factsheets/householdproducts.html, Accessed May 25, 2013. Database: http://householdproducts.nlm.nih.gov/, Accessed May 25, 2013.

45. International Programme on Chemical Safety (IPCS), World Health Organization, http://www.who.int/ipcs/en/, Accessed May 25, 2013.

46. Harper, Stacy. Nanomaterial-Biological Interactions Knowledgebase (NBIK), http://nbi.oregonstate.edu/ , Accessed May 25, 2013.

47. ACToR – Aggregated Computational Toxicology Resource, Environmental Protection Agency, http://actor.epa.gov/actor/faces/ACToRHome.jsp, Accessed May 25, 2013.

48. ChemSpider; the free chemical database, Royal Society of Chemistry, http://www.chemspider.com , Accessed May 25, 2013.

49. ChemSpider; the free chemical database – Data Sources, Royal Society of Chemistry, http://www.chemspider.com/DataSources.aspx, Accessed May 25, 2013.

50. ChemSpider; the free chemical database – Getting Started, Royal Society of Chemistry, http://www.chemspider.com/GettingStarted. aspx, Accessed May 25, 2013.
51. Hazard Information for Science of Synthesis, Thieme Medical Publishers, http://www.thieme-chemistry.com/fileadmin/Thieme/ SoS/pdf/hazard2.pdf, Accessed May 25, 2013.
52. *Exposure Science in the 21st Century: A Vision and a Strategy.* National Research Council, 2012, http://www.nap.edu/openbook. php?record_id=13507&page=1, Accessed May 25, 2013.
53. ChemHat.org – Chemical Hazard and Alternatives Toolbox, Communications Workers of America and BlueGreen Alliance, http://www.chemhat.org/, Accessed May 25, 2013.
54. National Oceanic and Atmospheric Administration (NOAA) Office of Response and Restoration, *Chemical Reactivity Worksheet*, http://response.restoration.noaa.gov/chemaids/react.html, Accessed May 25, 2013.

CHAPTER 8

Searching For Polymers

DONNA T. WRUBLEWSKI

California Institute of Technology, 1200 E. California Blvd., MC 1-43,
Pasadena, CA 91125, US
Email: dtwrublewski@library.caltech.edu

8.1 INTRODUCTION

A polymer is broadly defined as a substance made up of many linked atoms or smaller molecules. Unlike small molecules, polymer properties can vary considerably depending on the method of synthesis, measurement technique, and experimental conditions. As such, knowing additional information about the polymer of interest can be helpful when searching for specific information. This chapter outlines some of this additional information to aid in searching primary literature, and also includes references to secondary compendia where possible.

Section 8.2 covers polymer structure nomenclature and searching; although progress is being made towards a more uniform naming system, as with other areas of chemistry, there are colloquial as well as systematic names in use currently and particularly in older literature. Section 8.3 briefly describes common methods of polymer synthesis, as different synthetic methods applied to the same monomer can produce polymers with significantly different properties. Methods for characterization of polymer structure are discussed in Section 8.4, as different methods may yield different values for a property. Sections 8.5 and 8.6 describe some of the most common thermophysical and

Chemical Information for Chemists: A Primer
Edited by Judith N. Currano and Dana L. Roth
© The Royal Society of Chemistry 2014
Published by the Royal Society of Chemistry, www.rsc.org

mechanical properties of interest, respectively. Concluding remarks are offered in Section 8.7, along with a general bibliography for further background information.

8.2 POLYMER NOMENCLATURE AND STRUCTURE

In addition to standard text searching, chemical literature can also be searched *via* chemical structure. Basic text name searching will be discussed in Section 8.2.1 and structure-based searching is covered in Section 8.2.2.

8.2.1 Text Name Searching

There are three types of nomenclature currently in use for polymers: formal methodic scientific nomenclature systems, common colloquial names rooted in history, and commercial trade names. A thorough search should include as many names as possible, particularly when dealing with historic literature.

Historically, two classes of nomenclature systems have existed for the naming of polymers: source-based and structure-based. Overviews of each system can be found in several reference books, including the *Polymer Handbook*[1] and the most recent *CRC Handbook*,[2] as well as in introductory-level polymer chemistry texts such as that by Odian.[3] Source-based naming derives the polymer name from the monomer or monomers that form the polymer. Structure-based names are determined by examining the final structure of the polymer and identifying a repeating subunit. The International Union of Pure and Applied Chemistry recently published a compendium describing guidelines for a broad range of polymer terminology including nomenclature,[4] and recommend a structure-based nomenclature system for linear homopolymers that names polymers according to the smallest repeating subunit. However, for very common commercial polymers, this system is not expected to supplant established names. One example of this is polyethylene, so named for its formation from the monomer ethylene. The most basic structural repeat unit is that of the methylene group ($-CH_2-$), so IUPAC rules dictate the name should be polymethylene. Searching for polymethylene will yield very different results than for polyethylene; the former should be included to ensure older historical literature is retrieved. It is recommended to search for the IUPAC name in conjunction with the common name, should one exist.

Common established polymer names are frequently made up of several words, and the same name can often be written in more than one way. An example of this would be the terms *poly(ethylene oxide)* versus *polyethylene oxide*. In general discussion, parentheses should be used to write the polymer name when the monomer name contains multiple words or contains substituent position numbers.[5] However, when searching, it is important to know the specific database parameter, particularly if it is one with broad or non-specific subject coverage. For example, Thompson Reuters' Web of Knowledge platform executes Boolean expressions inside parentheses first.[6] For a Topic search with this example, parentheses are ignored and in the absence of a Boolean operator, *poly(ethylene oxide)* is searched as *poly* AND *ethylene* AND *oxide* and retrieves over 18,000 records. Similarly, *polyethylene oxide* is searched as *polyethylene* AND *oxide*, retrieving over 7,500 records. The first set will obviously include irrelevant records, while the second will likely be missing some relevant articles. Possible strategies for optimizing search results would include:

(1) Searching for "polyethylene oxide" OR "poly ethylene oxide" as phrases, which retrieves over 16,800 records from a Web of Knowledge Topic search, or, alternatively, over 5,300 from a Title search.
(2) Including additional terms, such as abbreviations or trade names, or other polymer-specific properties, thereby possibly eliminating references to small molecules.

Two general sources for polymer trade names and abbreviations are Section VIII in the *Polymer Handbook* (fourth edition)[7] and Appendix II in *Contemporary Polymer Chemistry*.[8] The *Polymer Data Handbook* is organized by polymer type, and lists trade names and abbreviations for many classes of polymers. The CAS Registry by default includes abbreviations and trade names in its comprehensive listings, and will be discussed later in this section.

Another type of compound name is encountered when dealing with copolymers. Source-based names can include multiple monomers, as well as the designation of the type of copolymer. The most common descriptors seen with copolymers are –co– (unspecified or unknown structure), –block– (monomers are in defined blocks within the polymer), and –alt– (monomer units alternate within the polymer), although others are becoming more common, along with more formal IUPAC designations for structures.[2] When searching for copolymers, it is helpful to not only search a polymer name (such as *poly(styrene-co-butadiene)*) but also phraseology such as "styrene butadiene copolymer" or "block copolymer" (if the type of copolymer structure is known). Trade names and abbreviations are also very common in the copolymer field, and can significantly enhance the relevance of searches if they are known.[7,8]

Terms relating to polymer stereochemistry can often be helpful in searching for information. Tacticity generally refers to the orientation of substituents on a (usually saturated) polymer backbone chain relative to the plane of the chain itself.[9] Substituents with periodic repetition located on the same side of the plane are referred to as isotactic, whereas if they are on opposite sides of the plane they are called syndiotactic. If a chain has no order to the placement of substituents relative to the plane, it is referred to as atactic. Usually these terms (or their abbreviations i-, s-, and a-) are used in conjunction with the polymer name, and can be used to narrow down a search if known. For unsaturated backbones, such as polyisoprene, another type of tacticity can also exist, that of transtactic and cistactic. For planar double bonds, *cis* refers to the polymer chain bonds being on the same side of the double bond, and *trans* if they are on opposite sides. Tacticity can have a significant effect on polymer properties, such as glass transition temperature and crystallinity, and thus should be included, if relevant, in searching for specific properties (*i.e.*, *"isotactic polypropylene"*).

8.2.2 Structure Searching

Small molecules, including monomers and some oligomers, can be searched by chemical structure in some databases, including the CAS Registry (accessible through the SciFinder or STN platforms),[10] Elsevier's Reaxys, NCBI's PubChem, and RSC's ChemSpider. Structures can be entered using text representation systems, such as the IUPAC International Chemical Identifier (InChI) or the Simplified Molecular-Input Line-Entry System (SMILES). Alternatively, on some platforms, structures can be drawn directly using embedded structure editors. Currently, there is no formal standard for polymer representation using SMILES notation,[11] however, development is ongoing in that area.[12] Also, development for InChI representation of polymers is ongoing but not standardized or widely adopted.[13] As such, there is limited support for polymer searching by full structure in any major database, but several do support searching by monomer structure. Examples of how to do this using Elsevier's Reaxys and the CAS Registry are in the following sections.

8.2.2.1 Reaxys. At the time of this writing, Reaxys has very limited retrieval of polymers using the "Structure and Properties" search facet, since it does not index polymers by structure. This includes patented and trade materials, as well as broad polymer groups. Some common polymers, such as polyethylene, can be searched using the "Generate structure from name" feature, however this will result in only the repeating unit (-CH2-CH2-) being identified and searched. More relevant, but again limited, results can be found by searching under the "Substances and Properties" function, selecting "Properties (Advanced)", then "Identification", then "Substance Identification", and populating the "Chemical Name" form with the prefix *poly* to generate a list of polymer names.

Another approach would be to use the "Reactions" function, and drawing or generating a monomer (or monomers) of interest and selecting "Starting material" under "Search as/by". Under "Conditions (Form-based)", the "Product name" form can be populated using the "starts with" option and the prefix *poly* to generate a list of index terms of possible polymer products. Of note here is that the Beilstein database, on which Reaxys is based, historically never indexed polymers, although the Gmelin database (also included in Reaxys) did include a limited number of inorganic polymers. It does appear though that, as with other sources, coverage is increasing due to indexing from journal articles and patents.

8.2.2.2 The CAS Registry Database. The CAS Registry is a specialized chemical substance database that, as of this writing, contains substance information, including names, structure, and property data, for over 70 million compounds, mostly small molecules and oligomers, each with a uniquely assigned CAS Registry Number (CAS RN).[10] An overview of the CAS Registry organization, as well as a list of CAS RN's for some common polymers, can be found in Section VIII of the *Polymer Handbook*.[14] Since the time of that publication, the coverage of the CAS Registry has grown exponentially and has become available on the SciFinder platform in addition to STN platform. Both platforms allow for structure searching by text or by drawing, but again are primarily limited to monomer structures. One advantage, however, of using a CAS RN to search is that unique chemical moieties have unique CAS RNs; chemicals with the same structure but different types of isomerism would have different CAS RNs, which can be helpful when searching for polymers with specific tacticity or activity. Knowing the CAS RN can also be of use when searching other resources that index them, including databases such as PubChem as well as vendor catalogs.

The SciFinder "Explore Substances" search facet allows searching of the CAS Registry by several fields. The three that will be discussed here in relation to polymers are "Substance Identifier", "Chemical Structure", and "Molecular Formula". The result of any search for a polymer will be a list of substances organized by CAS RN, and will include known alternate names (which can be helpful for text name searching), the molecular structure and CAS RN of any component monomers, and molecular property and spectral information for the polymer, if available, with references.

For common polymers, searching by name in the "Substance Identifier" field will usually identify the polymer. As in Reaxys, polymers with specific components (including many patented and trade materials, and broad groups such as polycarbonates) usually will not have structure information available. When using the "Chemical Structure" search facet, the structure of the monomer(s) of interest can be drawn, and the search can be limited to the "Polymers" Class. Polymers which cannot be found via a substance identifier search might be located if the component monomer(s) can be drawn and searched. The "Molecular Formula" field search allows searching by repeat unit, grouped by parentheses, followed by "x"; an example is *(C2H4)x* for polyethylene. For copolymer searches, include the formulae for each monomer separated by a period, such as *(C2H4.C2H4O)x* for co-polymers of ethene and ethylene oxide.

8.3 POLYMER SYNTHESIS

When searching for polymerization reactions, either in general or with respect to a specific polymer as a product, it is helpful to know the reactants, mechanism, and reaction conditions, as with any chemical reaction. Including the polymerization mechanism (chain growth or step growth; cationic or anionic; ring opening metathesis polymerization, also known as ROMP; *etc.*) in a search for a specific polymer and product characteristics such as target molar mass or tacticity can increase the relevance of search results. Overviews of common types of polymerization reactions can be found in many introductory-level polymer science texts. The synthesis of both general polymer types as well as industrially important materials can be found in the first two volumes of the work of Elias[15] as well as other references listed at the end of this chapter. When searching for polymers, particularly those that may be synthesized by more than one mechanism, including the polymer name and the desired polymerization method or reaction in a Boolean search strategy is recommended (*i.e.*, polystyrene, which can be made by either cationic or anionic mechanisms).

One common polymerization reaction parameter often desired is the rate constant. This value will depend heavily on the monomer(s), the type of polymerization, and the reaction conditions. A comprehensive list of reaction rates and reactivity ratios can be found in Section II of the *Polymer Handbook*.[16]

8.4   POLYMER STRUCTURE CHARACTERIZATION

As mentioned earlier, unlike small molecules, polymer molecules can have a range of properties, based not only on structure but also on differing molar masses with the same basic structure. This section will discuss the basics of molar mass properties and polymer structures, as well as characterization methods by which they are determined, since different methods can yield different results. As with synthesis information, it can be helpful to include the specific measurement technique when searching for a polymer so that the values obtained can be interpreted in context. Section 8.4.1 will discuss techniques and terminology related to molar mass measurements. Section 8.4.2 will briefly describe spectroscopic techniques frequently employed to determine polymer chemical structure. Physical structure characterization is treated in Section 8.4.3.

8.4.1   Molar Mass and Dispersity

In keeping with recent texts,[4,17] the terms "molar mass" and "dispersity" will be used here instead of molecular weight and polydispersity index, although the latter are still in common usage and prevalent in older literature. Molar mass is the weight of one mole of polymer molecules. Molecular weight is more correctly defined as the dimensionless relative molecular mass. The most common types of molar mass measured are the number-average (M_n), weight-average (M_w), and z-average (M_z). The dispersity of a polymer system is defined as the ratio of the weight-average molar mass to the number-average molar mass.

Different measurement techniques may yield different values for these properties. Techniques based on size exclusion chromatography (SEC), such as gel permeation chromatography (GPC), can determine the entire molar mass distribution of a polymer but require comparison with a well-characterized standard material.[18] Static light scattering (sometimes abbreviated as SLS) can be an effective way to determine an absolute value for M_w, as well as other molecular properties, and dynamic light scattering (DLS) can be used to determine properties such as diffusion coefficient and hydrodynamic radius.[19] Mass spectroscopy is another method of characterizing structure and molar mass, but until recently was only applicable to low molar mass compounds.[20]

Matrix-assisted laser desorption ionization time-of-flight (MALDI-TOF) is a method applying this technique to high molecular weight compounds, and is based on polymer chains associating with ions in the vapor phase. Measuring the time of flight of the polymer/ion complex through an applied electromagnetic field will give a distribution of travel times, with heavier species taking longer to travel through the detector.[21]

Because of the relationship between measured properties and molar mass, it is important to have a general idea of the molar mass range of interest when searching for property data. Likewise, when evaluating property data, it is important to be aware of the molar mass range and measurement technique for which it has been measured. When searching the primary literature with Boolean operators, molar mass ranges and measurement techniques (full names and acronyms) should be included in the search string construction, keeping in mind the terms used in older literature discussed previously. Of the secondary sources available, the *Polymer Data Handbook*[22] specifically compiles molar masses by trade name within polymer classes and types. For light scattering, a listing of scattering factors can be found in Section VII of the *Polymer Handbook*.[14]

8.4.2 Spectral Analyses

Many types of spectral analyses are useful for elucidating polymer structure. Brief descriptions will be given here for infrared spectroscopy, Raman spectroscopy, and nuclear magnetic resonance spectroscopy, along with pertinent search terms and recommended sources.

Infrared (IR) and Raman spectroscopy are both based on the interaction of infrared radiation with chemical bonds. IR spectroscopy measures the absorption of specific frequencies in the IR spectrum by chemical bonds that produce a change in dipole moment under irradiation.[23] IR spectra are used primarily for identification of substances by measuring the characteristic frequencies associated with specific molecular motions. Many current machines, usually of the Fourier Transform (FT-IR) type, come with software containing a database of established spectra and will automatically identify samples analyzed. Another type of IR is attenuated total reflectance (ATR-IR), which allows for surface characterization.

Raman spectroscopy is based on measuring the frequency change of inelastically scattered light.[24] Like IR, Raman is used to measure specific chemical bond energy properties, but it can also detect morphological changes as well. Less specific sample preparation is needed than with IR, so it can be employed in more environments, such as in-situ monitoring of polymerization reactions.

Nuclear magnetic resonance (NMR) spectroscopy is based on the principle of magnetic spin and measures radiation absorption (in the range of 1–500MHz) relative to a reference compound.[25] NMR spectroscopy usually measures ^{1}H or ^{13}C absorptions of polymers in solution, and can be used to determine more detailed microstructural information, such as tacticity, than that obtained *via* IR or Raman. Many types of advanced analysis techniques exist for NMR, including two-dimensional correlation spectroscopy (2D COSY) among others. Including these types of analysis techniques can be helpful in searching for information.

One of the most extensive collections of Raman, IR, NMR, and other spectra is the Sadtler Standard Spectra collection. Still available in many libraries in print, the collection was purchased from Sadtler Research Labs in 1978 by Bio-Rad, which currently maintains the electronic version of these spectral databases.[26,27] The print collection can be searched by indexes if information such as name, molecular formula, or chemical type is known. Recently, some spectra from the Bio-Rad electronic collection have been made available as part of substances' CAS Registry records, accessible via SciFinder and STN, and can be found by searching the CAS Registry for specific substances as discussed previously. While SciFinder does not generally index spectral data if it is used purely for characterization purposes, checking the full text of the original articles dealing with the synthesis of a compound may provide said data. Reaxys also allows searching by spectral availability and provides links to primary source data.

Some spectral data, such as NMR shifts and IR absorption frequencies, can be found for certain polymers in the *Polymer Data Handbook*[22] and in Section V of the *Polymer Handbook*.[16] Absorption tables for common chemical moieties exist in organic and polymer chemistry textbooks, such as that by Silverstein.[28] Tables of standard values for chemical shifts can be found in many textbooks and are often incorporated into equipment software and analysis. As before, when searching the primary literature it is helpful to include the specific type of technique or spectra desired (*i.e.*, ^{13}NMR as opposed to simply NMR).

8.4.3 X-Ray Diffraction and Scattering Analysis of Structure

X-ray Diffraction (XRD), Wide Angle X-ray Scattering (WAXS), and Small Angle X-ray Scattering (SAXS) are used to elucidate crystalline and semicrystalline phases in polymers that are capable of crystallization.[29] Technically, diffraction refers to the scattering pattern produced by the interaction of X-rays with regular features (such as crystals) in molecules. The size of a crystalline unit cell can be calculated from measuring the spacing of features in these patterns. Crystal symmetry systems, both plane and space groups, are compiled by the International Union of Crystallography in the International Tables.[30]

Scattering techniques measure scattered radiation (X-ray) intensity as a function of scattering angle.[31] WAXS studies can measure the spacing between individual chains in ordered regions and from this the degree of crystallinity and density can be calculated. SAXS can yield information about slightly larger molecular structures, including crystal thickness and periodicity. Crystallinity can be highly system-dependent, so it is important when searching the primary literature to include conditions such as molar mass, density, tacticity, measurement technique, and composition, if known (particularly for copolymers).

Crystallographic data, including space group information and unit cell parameters, for some polymers can be found in Section V of the *Polymer Handbook*[16] and the *Polymer Data Handbook*.[22] The Cambridge Structural Database and WebCSD[32] also include crystallographic data for some polymers, but is only searchable through a "Text/Numeric Search" and using "Compound Name", or entering the molecular formula in an "All Text" search. In SciFinder, when retrieving references from a "Substance Detail", it is possible to limit to those references containing "Crystal Structure" information.

8.5 POLYMER THERMOPHYSICAL PROPERTIES

This section describes a few of the most common thermophysical properties of polymers for which data is often desired. Many electronic resources, including SciFinder, Reaxys, and the *CRC Handbook of Chemistry and Physics*, have recently incorporated the ability to search by a specific value or range of values for a property of interest, instead of by substance. This includes some of the previously discussed properties such as spectra, and will include the properties discussed in Section 8.6. In general, searching by property value alone will yield relevant results provided the chosen database has an adequate amount of indexed property data and contains substances with the desired property value. If searching by property value fails to yield relevant information, searching primary literature by substance and property name is recommended.

8.5.1 Glass Transition, Melting, and Decomposition Temperatures

The types of thermophysical properties available for any given polymer will depend on its structure. Polymers can have the same chemical structure, but have significantly different physical structures depending on their formation conditions and processing conditions. One of the most important properties is the glass transition temperature (T_g). Most small molecules do not possess this transition due to significant order in the solid state – therefore, usually only a melting temperature is reported. Furthermore, for highly regular structures, this temperature will not vary by measurement method. Amorphous polymers will usually only display a glass transition temperature. Semi-crystalline or crystalline polymers may exhibit T_g in addition to T_m.

As mentioned previously, XRD is one way to determine crystallinity of polymers. Differential Scanning Calorimetry (DSC) is another technique used to measure crystallinity as well as glass and melting transitions, and heat capacity. It is based on measuring heat flow into (or temperature change of) a sample compared to a reference material.[33] The main experimental parameter in DSC is the rate of change in the heating element, as this affects the measured transition temperatures, and should be noted when searching for values listed above measured with this technique.

Glass transition and melting values for common polymers can be found in most introductory texts, the *Polymer Data Handbook*,[22] Section V of the *Polymer Handbook*,[16] and Section 13 of the *CRC Handbook of Chemistry and Physics*.[34] Attention should be paid to the method by which the value was determined.

The decomposition temperature of a polymer may be needed for specific applications, or for determining other experimental parameters. Usually this is measured *via* thermogravimetric analysis (TGA), which measures mass loss as a function of temperature. TGA experiments can be carried out at different heating rates in a variety of gaseous environments such as air, pure oxygen, or pure nitrogen, with or without humidity. These conditions should be included, if known, when searching for decomposition information in the primary literature. A compilation of common polymer decomposition temperatures and corresponding decomposition products is given in Section II of the *Polymer Handbook*.[35]

8.5.2 Polymer Solubility and Miscibility

Polymer solubility has implications for many experimental applications, including synthesis and characterization. When designing an experiment that requires a polymer solution, a list of common solvents can be found in Section 13 of the *CRC Handbook of Chemistry and Physics*.[34] More data for solvents can also be found in Section III of the *Polymer Handbook*.[16] Potential solvents can be used in conjunction with the polymer of interest to search the primary literature. Where water is the solvent of interest, the *CRC Handbook of Thermodynamic Data of Aqueous Polymer Solutions* provides solubility and other thermodynamic data for a number of polymers,[36] and the same is provided for copolymer solutions in a companion volume.[37]

Solubility is also directly related to polymer solution phase behavior, particularly in terms of upper critical solution temperatures (UCST) or lower critical solution temperatures (LCST). Again, some values are tabulated,[38] but including UCST or LCST terms in searches will help narrow down the desired phase properties.

8.6 POLYMER MECHANICAL AND ENGINEERING PROPERTIES

As noted for Section 8.5, several electronic databases allow for searching by property value. Again, should searching by property value provide inadequate information, searching the primary literature by substance and property name is recommended. Of particular note here is the common use of commercial materials for many polymer engineering studies, and a good source of property data for common commercially available materials is MatWeb,[39] where data is usually provided from commercial suppliers and MSDS sheets.

8.6.1 Viscoelastic Behavior

Due to the long chain molecular structure of polymers, their mechanical behavior is referred to as viscoelastic – a combination between that of a viscous fluid and an elastic solid. The mechanical properties of polymers are related by what is commonly called the *time temperature superposition (TTS) principle*.[40] In essence, this states that properties measured at a high rate are equivalent to properties measured at a low temperature, and *vice versa*. This is due to the effect of rate and temperature on molecular mobility. At low temperatures, there is less thermal energy to effect molecular rearrangement, and at high rates there is less time. At high temperatures, there is more thermal energy to effect molecular rearrangement, and at low rates there is more time. As such, when searching for particular property values, it is important to know beforehand the experimental conditions of interest, and to be aware of the conditions under which reported values are obtained. The first of the following two sections covers low-rate testing (usually referred to as static or quasi-static conditions), and the second covers high-rate testing (referred to as dynamic conditions).

8.6.2 Static Testing

Common static mechanical properties of relevance include tensile (or Young's) modulus, compressive modulus, shear modulus, ultimate tensile strength, and Poisson's ratio. Occasionally the term *compliance* is seen instead of modulus, and is the reciprocal of the modulus. Temperature of measurement, particularly with respect to the glass transition temperature of the material, is vital, as properties can change drastically even over the course of a fraction of a degree if in proximity to T_g. Rate of testing is also important, although properties tend not to vary much as long as the rate can be considered static or quasi-static.

Several secondary sources already mentioned provide mechanical property data for many polymers. Section 13 of the *CRC Handbook of Chemistry and Physics* provides strength data for nineteen common industrial polymers.[34] Section V of the *Polymer Handbook*[16] as well as the *Polymer Data Handbook*[22] both provide mechanical property data for many polymers, but are more easily navigable by polymer name than by property value.

When searching primary literature, properties such as molecular weight, temperature, and rate of testing should be included along with polymer name or structure. However, these properties can vary significantly by molar mass and crystallinity, so these additional parameters should be included if known.

8.6.3 Dynamic Testing

Dynamic mechanical properties, such as storage modulus, loss modulus, tan delta (the ratio of loss modulus to storage modulus) and dynamic viscosity are specific to both temperature and frequency of measurement, as well as to the intrinsic polymer properties. They can also vary by instrumentation used; a Dynamic Mechanical Analyzer (DMA) may give different data than a rheometer. When searching for or evaluating dynamic data, the experimental conditions (such as temperature or shearing rate, and frequency or frequency spectrum) should be well noted. Dynamic data is reported for some polymers in the *Polymer Data Handbook*.[22]

8.7 GENERAL POLYMER SCIENCE REFERENCE SOURCES BIBLIOGRAPHY

The following sources include the resources mentioned in previous sections, along with other sources that may be of use for finding polymer properties.

H. R. Allcock, F. W. Lampe and J. E. Mark, *Contemporary Polymer Chemistry*, 3rd Edition, Pearson, Upper Saddle River, 2003.

CAS Registry, American Chemical Society, http://www.cas.org/content/chemical-substances.

A. F. M. Barton, *CRC Handbook of Polymer-Liquid Interaction Parameters and Solubility Parameters*, CRC Press, Boca Raton, 1990.

The PubChem Project, Biotechnology Information, National Center For, http://pubchem.ncbi.nlm.nih.gov.

E. A. Grulke, A. Abe and D. R. Bloch, *Polymer Handbook*, 4th Edition, J. Brandrup, E. H. Immergut, John Wiley & Sons, New York, 1999.

Cambridge Structural Database (CSD), Cambridge Crystallographic Data Centre (CCDC), http://www.ccdc.cam.ac.uk/.

H. G. Elias, *Macromolecules*, 3rd Edition, Wiley-VCH, Weinheim, 2005.

Reaxys, Elsevier Properties SA, http://www.reaxys.com.

W. M. Haynes, *CRC Handbook of Chemistry and Physics*, 92nd Edition, CRC Press/Taylor and Francis, Boca Raton, FL, 2012.

P. C. Hiemenz and T. Lodge, *Polymer Chemistry*, 2nd Edition, CRC Press, Boca Raton, 2007.

International Tables for Crystallography, International Union of Crystallography, http://it.iucr.org.

A. D. Jenkins, P. Kratochvil, R. G. Jones, E. S. Wilks, W. V. Metanomski, J. Kahovec, M. Hess, R. Stepto and T. Kitayama, *Compendium of Polymer Terminology and Nomenclature: IUPAC Recommendations 2008*, The Royal Society of Chemistry, Cambridge, 2009.

MatWeb, MatWeb LLC, http://www.matweb.com.

J. E. Mark, *Polymer Data Handbook*, 2nd Edition, Oxford University Press, Oxford, 2009.

Polymer Database (PoLyInfo), National Institute for Materials Science (Japan), http://polymer.nims.go.jp/index_en.html.

G. G. Odian, *Principles of Polymerization*, 4th Edition, Wiley-Interscience, Hoboken, 2004.

Protein Data Bank, Research Collaboratory for Structural Bioinformatics (RCSB), http://www.rcsb.org.

R. M. Silverstein, D. J. Kiemle and F. X. Webster, *Spectrometric Identification of Organic Compounds*, 7th Edition, John Wiley & Sons, Hoboken, NJ, 2005.

L. H. Sperling, *Introduction to Physical Polymer Science*, 4th Edition, Wiley, Hoboken, NJ, 2006.

Polymers: A Properties Database, Taylor & Francis Group, http://http://www.polymersdatabase.com/.

Nucleic Acid Database (NDB), The State University of New Jersey, Rutgers, http://ndbserver.rutgers.edu/.

C. Wohlfarth, *CRC Handbook of Thermodynamic Data of Aqueous Polymer Solutions*, CRC Press, Boca Raton, 2004.

C. Wohlfarth, *CRC Handbook of Thermodynamic Data of Copolymer Solutions*, CRC Press, Boca Raton, 2001.

8.8 CONCLUSIONS

Searching for polymer information, like other types of chemical information, is challenged by both the changing nature of search interfaces and subject databases, as well as the evolution of the nomenclature and structure representation itself. An added complication is the dependence of polymer properties on both physical (in addition to chemical) structure as well as the method of measurement. Secondary sources and data compilations should be consulted whenever possible, but attention should be paid to the caveats above when reporting data. When searching the primary literature, knowing specific information of interest, such as substance properties or characterization techniques, can facilitate finding relevant information.

ACKNOWLEDGEMENTS

The author is greatly indebted to the librarians and staff at the Marston Science Library at the University of Florida for providing the time and support to allow the completion of this work.

REFERENCES

1. W. V. Metanomski, *Polymer Handbook*, ed. J. Brandrup, E. H. Immergut, E. A. Grulke, A. Abe and D. R. Bloch, John Wiley & Sons, New York, 4th, 1999, Nomenclature, pp. I/1–I/12.
2. R. B. Fox and E. S. Wilks, *CRC Handbook of Chemistry and Physics*, ed. W. M. Haynes, CRC Press/Taylor and Francis, Boca Raton, FL, 92nd, 2012, Nomenclature for Organic Polymers, pp. 13-5–13-8.
3. G. G. Odian, *Principles of Polymerization*, Wiley-Interscience, Hoboken, 4th, 2004, p. 832.
4. A. D. Jenkins, P. Kratochvil, R. G. Jones, E. S. Wilks, W. V. Metanomski, J. Kahovec, M. Hess, R. Stepto and T. Kitayama, *Compendium of Polymer Terminology and Nomenclature: IUPAC*

Recommendations 2008, The Royal Society of Chemistry, Cambridge, 2009, p. 464.

5. H. R. Allcock, F. W. Lampe and J. E. Mark, *Contemporary Polymer Chemistry*, Pearson, Upper Saddle River, 3rd, 2003, Appendix I Polymer Nomenclature, pp. 774–778.
6. Thompson Reuters, Web of Science Help - Search Operators, http://images.webofknowledge.com/WOK45/help/WOS/ht_operators.html (last accessed February 2013).
7. H. Elias, *Polymer Handbook*, ed. J. Brandrup, E. H. Immergut, E. A. Grulke, A. Abe and D. R. Bloch, John Wiley & Sons, New York, 4th, 1999, Abbreviations for Thermoplastics, Thermosets, Fibers, Elastomers, and Additives, pp. VIII/1–VIII/24.
8. H. R. Allcock, F. W. Lampe and J. E. Mark, *Contemporary Polymer Chemistry*, Pearson, Upper Saddle River, 3rd, 2003, Appendix II Properties and Uses of Selected Polymers, pp. 779–790.
9. G. G. Odian, *Principles of Polymerization*, Wiley-Interscience, Hoboken, NJ, 4th, 2004, Principles of Polymerization, pp. 620–637.
10. American Chemical Society, CAS Registry, http://www.cas.org/content/chemical-substances (last accessed February 2013).
11. D. Weininger, SMILES, *J. Chem. Inf. Comput. Sci.*, 1988, **28**, 31–36 (DOI:10.1021/ci00057a005).
12. A. Drefahl, *J. Cheminf.*, 2011, **3**, 1 (DOI:10.1186/1758-2946-3-1).
13. N. Day and P. Murray-Rust, The Unofficial InChI FAQ Section 4.15, http://www.inchi-trust.org/fileadmin/user_upload/html/inchifaq/inchi-faq.html#4.15 (last accessed February 2013).
14. E. F. Casassa, *Polymer Handbook*, ed. J. Brandrup, E. H. Immergut, E. A. Grulke, A. Abe and D. R. Bloch, John Wiley & Sons, New York, 4th, 1999, Particle Scattering Factors in Rayleigh Scattering, pp. VII/629–VII/636.
15. H. G. Elias, *Macromolecules*, Wiley-VCH, Weinheim, 3rd, 2005.
16. J. Brandrup, E. H. Immergut, E. A. Grulke, A. Abe and D. R. Bloch, *Polymer Handbook*, John Wiley & Sons, New York, 1999.
17. R. J. Young and P. A. Lovell, *Introduction to Polymers*, CRC Press, Boca Raton, 3rd, 2011, 1.3.2 Molar Mass Averages, p. 13.
18. R. J. Young and P. A. Lovell, *Introduction to Polymers*, CRC Press, Boca Raton, 3rd, 2011, 14.3 Gel Permeation Chromatography, pp. 318–323.
19. R. J. Young and P. A. Lovell, *Introduction to Polymers*, CRC Press, Boca Raton, 3rd, 2011, 12 Scattering Methods, pp. 281–297.
20. H. R. Allcock, F. W. Lampe and J. E. Mark, *Contemporary Polymer Chemistry*, Pearson, Upper Saddle River, 3rd, 2003, p. 432.

21. R. J. Young and P. A. Lovell, *Introduction to Polymers*, CRC Press, Boca Raton, 3rd, 2011, 14.5.1 Mass Spectra of Polymers, pp. 332–341.
22. J. E. Mark, *Polymer Data Handbook*, Oxford University Press, Oxford, *2nd*, 2009, p. 1264.
23. R. J. Young and P. A. Lovell, *Introduction to Polymers*, CRC Press, Boca Raton, 3rd, 2011, 15.4 Infrared Spectroscopy, pp. 351–355.
24. R. J. Young and P. A. Lovell, *Introduction to Polymers*, CRC Press, Boca Raton, 3rd, 2011, 15.5 Raman Spectroscopy, pp. 355–358.
25. R. J. Young and P. A. Lovell, *Introduction to Polymers*, CRC Press, Boca Raton, 3rd, 2011, 15.6 Nuclear Magnetic Resonance Spectroscopy, pp. 358–374.
26. L. Wang, *Chem. Eng. News*, 2008, **86**, 43–47.
27. Bio-Rad Laboratories Inc., Spectral Databases: Overview, http://www.bio-rad.com/evportal/en/US/INF/Category/203359/Spectral-Databases (last accessed February 2013).
28. R. M. Silverstein, D. J. Kiemle and F. X. Webster, *Spectrometric Identification of Organic Compounds*, John Wiley & Sons, Hoboken, NJ, 7th, 2005, p. 512.
29. R. J. Young and P. A. Lovell, *Introduction to Polymers*, CRC Press, Boca Raton, 3rd, 2011, 17 The Crystalline State, pp. 399–421.
30. International Union of Crystallography, International Tables for Crystallography, http://it.iucr.org (last accessed February 2013).
31. L. H. Sperling, *Introduction to Physical Polymer Science*, Wiley, Hoboken, NJ, 4th, 2006, 5.2.2.2 Electron and X-Ray Diffraction, pp. 207–208.
32. Cambridge Crystallographic Data Centre (CCDC), Cambridge Structural Database (CSD), http://www.ccdc.cam.ac.uk/ (last accessed February 2013).
33. R. J. Young and P. A. Lovell, *Introduction to Polymers*, CRC Press, Boca Raton, 3rd, 2011, 17.8.1 Differential Scanning Calorimetry, pp. 435–440.
34. W. M. Haynes, *CRC Handbook of Chemistry and Physics*, CRC Press/Taylor and Francis, Boca Raton, FL, 92nd, 2012.
35. J. Liggat, in *Polymer Handbook*, ed. J. Brandrup, E. H. Immergut, E. A. Grulke, A. Abe and D. R. Bloch, John Wiley & Sons, New York, 4th, 1999, Products of Thermal Degradation of Polymers, pp. II/451–II/475.
36. C. Wohlfarth, *CRC Handbook of Thermodynamic Data of Aqueous Polymer Solutions*, CRC Press, Boca Raton, 2004, p. 520.

37. C. Wohlfarth, *CRC Handbook of Thermodynamic Data of Copolymer Solutions*, CRC Press, Boca Raton, 2001, p. 520.
38. C. Wohlfarth, *CRC Handbook of Chemistry and Physics*, ed. W. M. Haynes, CRC Press/Taylor and Francis, Boca Raton, FL, 92nd, 2012, Upper Critical (UCST) and Lower Critical (LCST) Solution Temperatures for Binary Polymer Solutions, pp. 13-26–13-43.
39. MatWeb LLC., *MatWeb*, http://www.matweb.com (last accessed February 2013).
40. J. D. Ferry, *Viscoelastic Properties of Polymers*, Wiley, New York, 3rd, 1980, Viscoelastic Properties of Polymers, pp. 1–32.

CHAPTER 9

Reaction Searching

JUDITH N. CURRANO

Chemistry Library, University of Pennsylvania, 231 S. 34th St., 5th Floor, Philadelphia, PA 19104-6323, US
Email: currano@pobox.upenn.edu

9.1 INTRODUCTION

If you have done any form of synthetic chemistry, chances are that you have performed at least one graphical reaction search. You draw the reactant, the product, or both participants into the structure editor of an electronic resource that searches reaction schema, click the "search" button, sit back, and wait for results to fill your screen. If you are searching for an approach to the synthesis of a novel molecule, you probably employed substructures in your reactant, your product, or both. This chapter will examine general techniques in graphical reaction searching before presenting various resources that allow graphical or textual methods of finding reaction information and then will review several different sources of information about reactions and reagents. We will focus mainly on the reactions of organic and organometallic substances, as those are the ones treated most fully by the available reaction search tools.

There is a wide variety of information sources available, and, just because we do not treat them in this chapter, you should not assume that they are not worthy of mention. In general, we will avoid the use of screen shots from the tools that we discuss, as database interfaces change frequently. We will also avoid specific mention of tools and

Chemical Information for Chemists: A Primer
Edited by Judith N. Currano and Dana L. Roth
© The Royal Society of Chemistry 2014
Published by the Royal Society of Chemistry, www.rsc.org

clicks to use within these resources by employing generic structures and schema instead. Many search examples will employ Chemical Abstracts Service's CASREACT database and Elsevier's Reaxys® database, as these are two of the largest graphical reaction databases available today. However, the search techniques demonstrated should be transferrable to whatever systems are available at your home institution. The resources presented in Section 9.3 are demonstrative of the types of literature that they represent, but, once again, they are not the only quality tools available in their genres.

9.2 TECHNIQUES OF REACTION SEARCHING

9.2.1 Types of Graphical Reaction Search

There are two basic types of graphical reaction search that you can perform: partial and complete. They are good for different types of information need, and they employ slightly different search strategies, some of which can be used in either type of search, and others of which are unique to the type in which we present them. We will begin with partial reaction searching, in which you draw either the reactant or the product, and proceed to complete reaction searching, in which both sides of the reaction are drawn.

9.2.1.1 Partial Reaction Searches. For the purposes of this book, we will define a partial reaction search as one in which either the reactant or the product is specified and the other participant is left open. This technique is useful in several cases, listed below.

- You want to know how to synthesize a particular molecule or class of molecule, and you do not wish to limit your options by specifying a starting material.
- You want to know in which types of reaction a particular substance or class of substance participates.
- You want to know which types of reaction a particular substance or class of substance catalyzes.

Performing a partial reaction search is a two- or three-step process, depending largely on the tool that you select. First, you need to identify the substance or type of substance that interests you. This could be through a substance search or a drawn structure or substructure search in the resource's reaction editor. Once you have identified the substance, you must indicate its role in the reaction. In many cases, this will be done by drawing a reaction arrow that points towards or away from the substance. In other cases, you may select the substance's role using a radial button in the main search interface or a right click of the mouse. Finally, if you are drawing your structure into a graphical reaction editor, you may mark bonds to be formed or broken in the course of the reaction. Example 1 will guide you through the steps of such a search.

Example 1   Performing a Partial Reaction Search

Assume that you are interested in locating all reactions in which a chlorine atom is attached at the para *position of a carbon-substituted benzene ring. You are not particularly interested in specifying a starting material, as you would like to find as many possible methods as possible. First, draw the substructure, including the chlorine atom, into your structure editor. Next, you will want to specify that the structure you have drawn is a product. The easiest way of doing this in most systems is to draw a reaction arrow pointing towards the structure, as in Figure 9.1.*

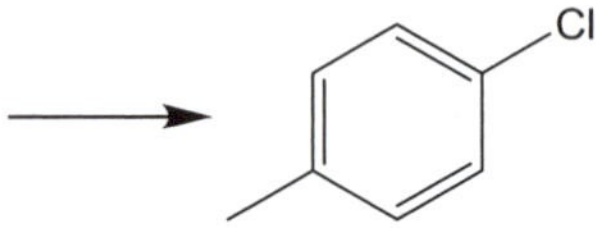

Figure 9.1   A partial reaction search for the synthesis of a chlorinated benzene ring.

In order to allow a variety of starting materials, it is best not to draw anything to the left of the arrow. The result will be any reaction in which a para *substituted, chlorinated benzene ring appears as a product. However, not all of resulting schema will display the synthesis of this particular segment of the product.*

Since you are interested in finding all reactions that form this structure by making the C-Cl bond, you need to build this piece of information into your search. Most reaction editors include a tool that allows you to mark a bond that is to be formed or broken in the course of the reaction. Use this tool to mark the bond between the chlorine atom and the carbon in the ring, as in Figure 9.2.

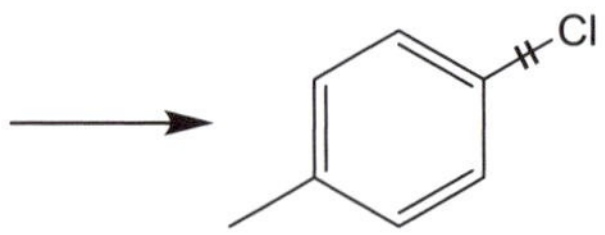

Figure 9.2   The bond between the carbon and chlorine atoms is designated as a bond to be formed or broken, using the hash tool in the resource's reaction editor. If used in a reactant, this indicates that the bond will be broken; when employed in a product, as in this case, the marked bond will be formed.

The resulting reactions will include a product with the drawn sub-structure, as well as forming the C-Cl bond of interest.

The "bonds to be formed or broken" tool can be used in a variety of ways, both in partial and in complete reaction searches. While Example 1 shows its use in a product to indicate a bond that must be formed in the reaction, it can also be used in a reactant to indicate that a bond is broken. For example, if you wish to break the carbon-oxygen bond in a furan, you can use a similar strategy, as shown in Figure 9.3.

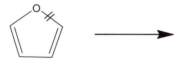

Figure 9.3   A reaction scheme to locate ring-breaking reactions in which the carbon-oxygen bond of a furan ring is disconnected.

9.2.1.2   Complete Reaction Searches. The partial reaction search strategy is helpful when you are looking for methods of forming or breaking a particular bond or when you want to do a comprehensive search for methods of forming a particular structure or substructure. However, it falls short when you wish to effect a transformation from a specific starting material to a product that you also specify. In this case, you will want to perform a complete reaction search. Again, there are three basic steps to follow. First, draw the structure or structures of the reactants and products that interest you. Then, designate the role of each in the reaction. The method of doing this will vary from one system to another, but the most universal way is to draw a reaction arrow from the reactants to the products. Finally, if desired, map atoms from reactants to products or indicate bonds to be formed or broken. This is not necessary for a successful reaction search, but it will improve precision, especially if your reactant and product structures are very general or if the same atom or group appears more than once in the substructure.

To demonstrate these steps, we will, once again, examine the case of the chlorination of a mono-substituted benzene ring.

Example 2   Performing a Complete Reaction Search

Assume that you want to add a chlorine atom to a monosubstituted benzene ring. You want to add the chlorine para *to the carbon substituent on the ring, and the position to which you plan to add the halogen is unsubstituted in the starting material (Figure 9.4).*

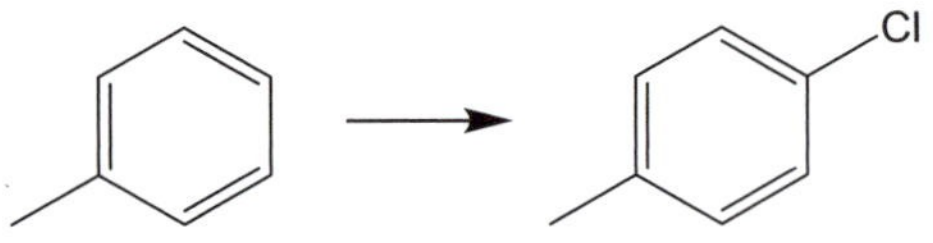

Figure 9.4   A reaction scheme to locate reactions that chlorinate a previously
unsubstituted position on a benzene ring.

*In this case, simply performing a partial reaction search in which the
carbon-chlorine bond is formed will be too general; it may yield examples
of halogen exchange reactions or other, undesired transformations. In this
case, it is best to do a complete reaction search, bearing in mind that, if it is
to be a substucture search, you must draw the hydrogen atom to ensure
that you retrieve the correct type of reaction (Figure 9.5).*

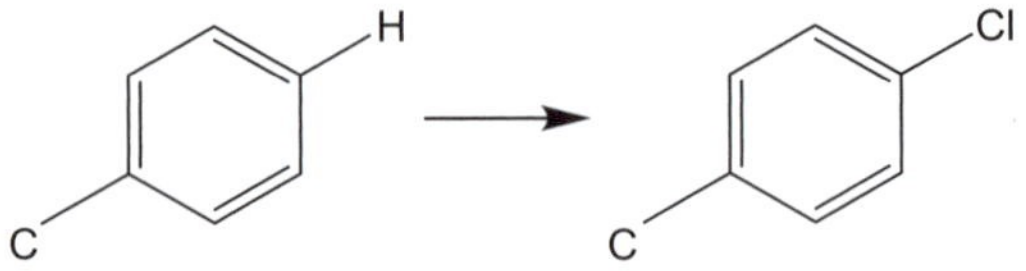

Figure 9.5   In order to more specifically retrieve the desired halogenation reactions, a
hydrogen atom is drawn in the para position of the reactant molecule.

*However, even this is not specific enough. Because of the generality of
the reactant substructure, it is still possible to locate many undesired
results, such as the transformation in Figure 9.6.*

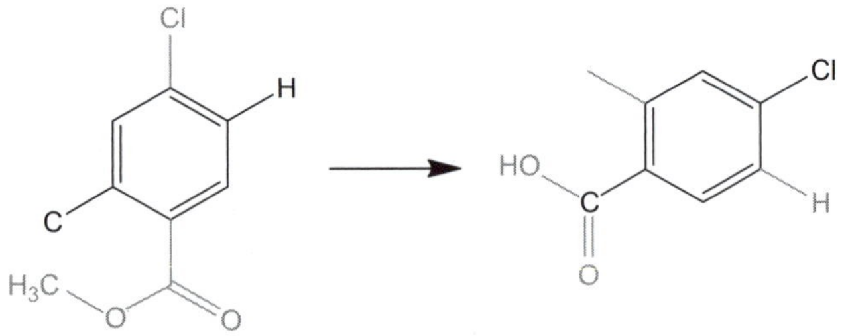

Figure 9.6   An undesired hit that could result from the reaction search in Figure 9.5.
The search substructure is shown in black, with other substitutents drawn
in gray. Note that the halogen appears in both the reactant and the product substructure, and that the reaction performed is actually the breaking
of the carbon-oxygen bond in the ester side chain.

*In this case, the desired structural elements appear in the reactant and
product, but the halogenation has not occurred. Therefore, it is necessary
to inform the system that, for the result to be acceptable, chemistry must
occur at one specific carbon center. This is done by mapping atoms from*

reactant to product with the resource's mapping tool, following to the vendor-supplied instructions. Most resources that allow graphical reaction searching have some kind of a mapping tool. The result will look like the scheme in Figure 9.7 and should retrieve the desired results.

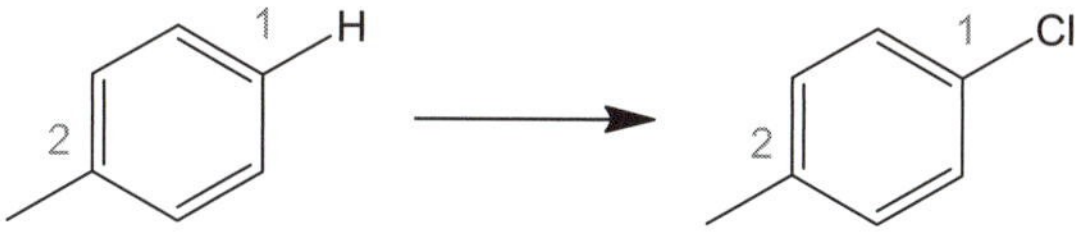

Figure 9.7 Mapping from reactant to product ensures that chemistry occurs at the desired centers. In this case, Carbon 1 will have a hydrogen attachment in the reactant but a chlorine attachment in the product, while Carbon 2 will have a carbon attachment in both molecules. This will ensure that the reactions retrieved are all chlorination reactions.

You should always attempt to make your structures as specific as possible, using hydrogen attachments or locking atoms to substitution everywhere appropriate; however, occasionally you will want more generality than these techniques afford you. When you have very general reactant and product substructures, not only should you mark the atom at which the chemistry will occur, but you can improve your results still further by marking one or more additional, distinctive atoms that appear in both structures. Bear in mind, however, that atom mapping, while it does improve the accuracy of your results, may slow your search considerably.

In some cases, depending on the nature of the reaction that you wish to locate, it may be helpful to indicate bonds that are formed and broken, sometimes in conjunction with atom mapping (Figure 9.8).

Figure 9.8 Marking a bond to be made or broken can supplement atom mapping for even greater precision. It is particularly helpful in large molecules, although even small schemes like this can benefit.

In a case like this, it would be overkill, but in a more complicated molecule, in which you may have several things happening at once, it could be very helpful.

9.2.2 Searching for General Methods and Transformations

When you are first approaching a tricky transformation or the synthesis of a complex molecule, it is helpful to examine a variety of methods of performing the reactions that you intend to employ. A graphical reaction search of the primary literature can quickly retrieve hundreds of reactions that perform your desired transformation, but, if you are not sure of the best way to proceed, it can be very daunting to sift through the many results for a promising preparation. In this case, a general methodology or review source is more helpful. These sources include enormous comprehensive sets, such as *Houben Weyl's Methoden der Organischen Chemie* and its successor, *Science of Synthesis*; smaller entities like *Comprehensive Organic Synthesis*; works focusing on reagents, like *The Encyclopedia of Reagents for Organic Synthesis*; and even single-volume works like *Larock's Comprehensive Organic Functional Group Transformations*, Greene's *Protective Groups in Organic Synthesis*, and Kocienski's *Protecting Groups*. These works and other more general resources for reaction information will be discussed in Section 9.3.

Both the benefit and the shortcomings of these types of works are the same: they are review sources, and you are viewing the literature through the eyes of an expert in the field. The authors and editors have assembled what they consider to be the most important methods of doing particular reactions in a single place. This allows them to perform some critical analysis on the various techniques, comparing the conditions, benefits, and downsides of all of the methods presented, and it presents you with some starting references and reaction schema that you can use when beginning your exploration of the primary literature. Frequently, methods will be displayed as a general reaction scheme containing R-groups to indicate substituents on the parent molecules and a table of the various substituents, conditions and results of each method. This makes this type of literature unique, in that it is represents one of the few easy ways to identify less-optimal methods of effecting a particular transformation. In addition, the authors and editors attempt to choose organizational schemata that emulate the thought process of a synthetic chemist, making it simple to identify a broad range of relevant information quickly.

Some of the review and methodology sources, such as *Science of Synthesis*, are structure searchable, and other sources rely on text, ring indices, and structure templates. When performing a structure search using the native interface of such a source, you want to be sure to keep it general; searching for a very specific reaction may yield no hits even though there are several methods that might work under the conditions you require. Of course, many of the chapters of methodology sources are indexed in *Chemical Abstracts*, and you can perform exact structure or reaction searches there.

Example 3   Locating a Method of Performing a Transformation Using Science of Synthesis

Assume that you wish to find a variety of methods of opening five-membered hetero rings containing nitrogen, where the disconnection occurs between the nitrogen and the adjacent carbon (Figure 9.9).

Figure 9.9   A ring-opening reaction, in which the marked bond is broken in order to open the ring.

You would like to know for which substituents on the ring this transformation would be particularly effective, as well as substrates that will not perform the desired ring-opening reaction; therefore, you decide to employ a review source, such as Science of Synthesis.

First, design a reaction substructure search, indicating all of your definite chemical requirements, but being careful not to allow the substructure to become too specific that you will retrieve no hits. A search for reactions that break the carbon-nitrogen ring bond is shown in Figure 9.10.

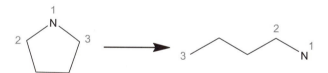

Figure 9.10   A general substructure that you would use to locate methods of performing the type of transformation in Figure 9.9. In the product, bonds between the nitrogen and its adjacent carbon and between the last two carbon atoms in the chain are marked as "chain only". Numbers indicate atom mapping, ensuring that the ring is actually broken.

Enter your search into the graphical search interface of the Science of Synthesis, *being sure to include any required catalysts or conditions for your reaction. Again, you wish to guard against adding too many limiting facts to draw the largest scope of information possible.*

Science of Synthesis *is organized by type of substance to be constructed, so, when you retrieve your hits, it is helpful to look at the volume and chapter in which each appears. By doing this, you can weed out undesired results quickly. The entries will indicate the appropriate conditions for this particular reaction, as well as any drawbacks to the technique, and references to the primary literature. While the reaction information in*

Science of Synthesis *usually provides a sufficient level of detail to repeat the reaction without referring to the primary literature, it can be useful to see the reaction in context and to backtrack through the references that the original author cited.*

9.2.3   Advanced Graphical Reaction Search Techniques

9.2.3.1   Forbidding Transformations at Certain Positions. It is very useful to be able to specify bonds that can be formed or broken in a reaction, but occasionally, you need exactly the opposite; you wish to specify that a particular site or functional group remains unchanged while you perform a transformation elsewhere in the molecule. The easiest way to do this is to use atom mapping. Assume, for example, that you wish to break a carbon-oxygen bond in an ester, while leaving the alcohol group adjacent to it unchanged, as in Figure 9.11.

Mapping atoms in both the ester and the alcohol will ensure that the ester is transformed but the alcohol is not. Note that the use of shortcuts is incompatible with atom mapping, so, if you wish to map atoms in a functional group you must draw all atoms individually in both the reactant and product (Figure 9.12).

This works beautifully when you know exactly how the substructure you want to change attaches to the substructure that you wish to retain; however, this is not always the case. Assume now that you wish to form an acid from an ester in the presence of a non-reacting alcohol, but you do not care where in the molecule that alcohol is attached. In a case like this, your method of setting and restricting reaction centers will depend on your choice of resource.

Setting a Non-reacting Functional Group

SciFinder® permits you to select functional groups from a list and give them a role of "non-reacting." You may do this either using the reaction editor, or in the general reaction search screen, and your functional

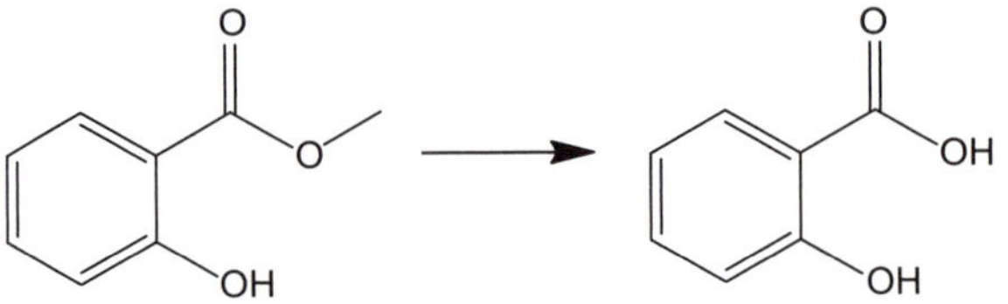

Figure 9.11   A transformation in which a carbon-oxygen bond is broken while leaving the adjacent alcohol unchanged.

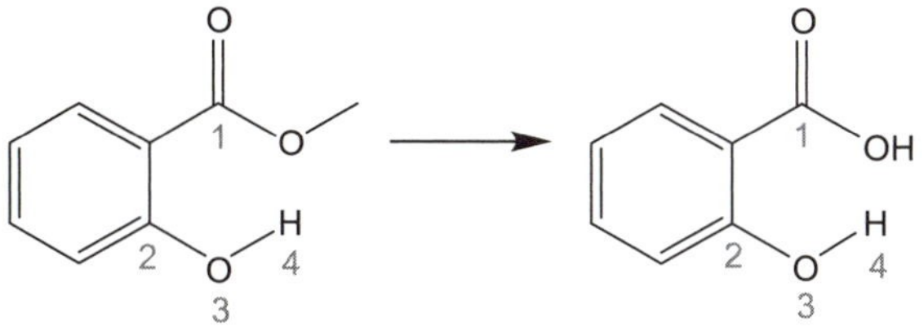

Figure 9.12   Mapping atoms in the ester ensures that the chemistry occurs at the desired position, while mapping the C-O-H in the reactant to the corresponding C-O-H in the product ensure that the alcohol remains the same.

group appears as a separate participant in the reaction. While this allows you to specify that the transformation is done in the presence of a non-reacting alcohol, it places no restrictions on where that alcohol is; it could be anywhere in the substrate, in a different reactant or reagent entirely, or even in the solvent. Therefore, although the search will return all reactions in which an alcohol in the substrate does not react, you may also get some irrelevant or undesired hits that you will have to filter by hand.

Refining Your Initial Reaction Search with a Second Search

This is a much more universal and slightly more precise way of setting the desired restriction. In this method, you draw a reaction search including only the desired transformation (Figure 9.13a). Once you have obtained your results, you can refine them by substructure to limit to only those that contain the functionality that remains unchanged from reactant to product (Figure 9.13b).

While this is also somewhat ambiguous in that the unchanged functionality could appear in a different reactant and product, it narrows the results somewhat and ensures that all desired results are present, while minimizing the number of undesired hits.

Indicating That a Center Is Not a Reaction Center

It is also possible, in systems like Reaxys®, to indicate that a bond between two atoms remains unchanged or that an atom is not a reaction center. This can be a useful technique, and it is a less time-consuming alternative to drawing a set of mapped reactions, as in the previous case.

9.2.3.2   Reaction Searching and Stereochemistry.   Stereochemistry, as in structure searching, can present challenges when performing a partial or complete reaction search in many resources. We will highlight some aspects of searching for reactions that include stereocenters in two of the larger reaction databases, CASREACT and Reaxys®, but

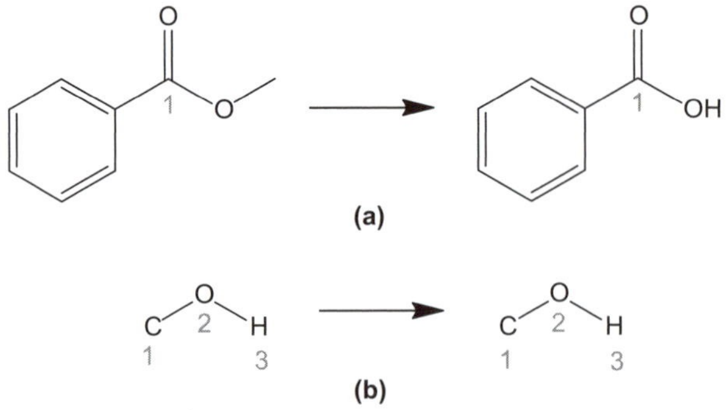

(a)

(b)

Figure 9.13   (a) A substructure reaction search for reactions in which the carbon-oxygen bond in an ester is broken. (b) A much more general substructure search, in which an alcohol group remains unchanged from reactant to product. This can be used to refine the search in Figure 9.13a, so as to ensure that the ester is transformed in the presence of a non-reacting alcohol.

it is important to refer to the help documentation of your chosen resource before embarking on such a search. In general, if you care greatly about stereochemistry in your reaction, you should, whenever possible, include it in the structures that you draw. However, be aware that, for a particular stereoconfiguration to be retrieved, it must be both published and indexed as having that stereochemistry. If the stereochemistry drawn in the original article is ambiguous or if the indexing of that article includes a reaction scheme that was drawn without stereochemistry, the reaction will not be retrieved in a search that makes use of explicitly drawn stereocenters. Therefore, if you desire comprehensive results, you may want to run your search twice, once including stereochemistry and once without stereo bonds drawn or with the stereochemistry option in your structure editor turned off.

9.2.3.2.1 Stereochemistry in CASREACT Reactions. The Chemical Abstracts Service's CASREACT database is one of the largest reaction databases in existence, but the reactants and products do not include stereochemistry. Therefore, if you draw stereocenters in your reactant or product, the system will inform you that they will be ignored and will proceed to retrieve all reactions that have the connections drawn, regardless of the stereochemical or geometric configuration. There are two ways that you can make your search slightly more specific.

Limit to Stereospecific Reactions

CASREACT has reactions indexed according to type. These types include catalyzed and non-catalyzed, which we will discuss later, regioselective, and stereoselective, along with other options. When using

SciFinder® to search CASREACT, you have the option of setting a limit of "stereoselective" at the beginning of the search or of limiting your results using this classification after you have viewed them. Although this will not limit your results to a specific configuration, it will remove all non-stereoselective reactions from your set, giving you fewer hits to review.

Searching for the Synthesis or Reactions of Known Substances or Classes of Substance

This method is a bit more cumbersome and involves performing several searches and combining search sets. It relies on the fact that substances, when indexed in references, are given reaction roles, so, you will be performing substance searches, as well as a reaction search. The following example illustrates a method of doing this.

Example 4   Searching for Stereochemistry in a Reaction Search in CASREACT, Using SciFinder®

Assume that you wish to locate the following type of reaction (Figure 9.14) in CASREACT, and you care greatly about stereochemistry.

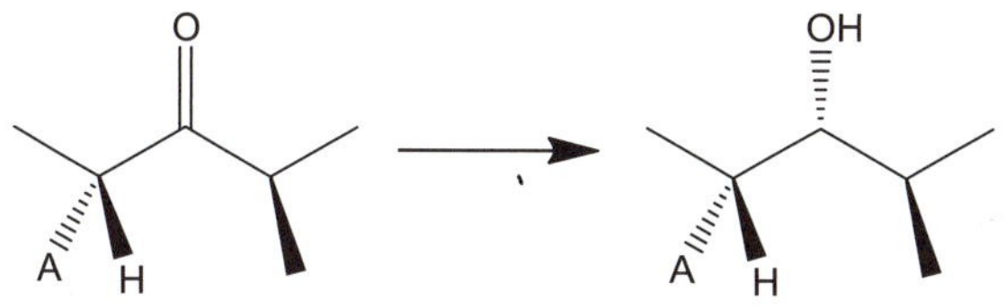

Figure 9.14   A stereoselective reaction that reduces a ketone to an alcohol. A represents any atom except hydrogen (see Section 5.2.3.1 for more information on using generics in substructures).

Step 1: Search for reactions that include this transformation, as shown in Figure 9.15. You need not draw the stereo bonds since the reaction search will ignore them anyway.

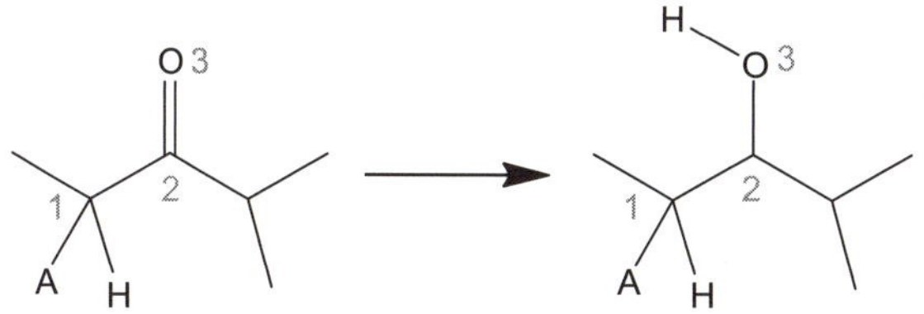

Figure 9.15   The reaction from Figure 9.14, drawn without stereochemistry. Note the atom mapping, used to increase search precision and weed out hits where the chemistry does not occur at the desired site.

Limit your search by reaction type, and choose "Stereoselective" from the list of options. When you have your answers, choose the option to "Get References.". This will retrieve all of the papers in which your answers appear. Save this answer set. Note that SciFinder® has an upper limit of results that it will save at a time, so, if your hit set exceeds this maximum, you will lose some hits when you save.

Step 2: Now, begin a second search, this time searching for substances that contain the product substructure from Figure 9.12, drawn with stereochemistry. From your list of substances, choose the option to get references, and, from the resulting menu, opt to get only those references in which the substances are prepared. You should now have a list of references that prepare your product.

Step 3: Select the "Combine Search Sets" option from the "Tools" menu, and combine your saved search for references including the complete reaction with your search for references that prepare the product. You will want to get the intersection of the two sets. This will give you all documents, in which a substance with your required stereochemistry is prepared and that include the desired reduction, although, in some cases, the reaction may not prepare the product with the given stereochemistry; instead, the reaction may produce a product with different stereochemistry, but your desired product is made using another reaction.

Step 4 (optional): If you have a large number of hits and wish to refine them further, save the references you got in Step 3, and begin a new substance search. This time, draw the reactant, as shown in Figure 9.12, including stereochemistry. From your list of substances, select the option to "Get References" and, from the resulting menu, get only those references in which the substances retrieved are reactants or reagents. Now combine your hit set with your saved results from Step 3, choosing the option to intersect the two sets. Your answers now will have your desired reactant as a reactant, will prepare your desired product, and will include the transformation that you want to find.

As you can see, the procedure is extremely time consuming, involving at least two searches and one combination of search sets. In addition, it is not always guaranteed to be successful, since, at the time of this writing, SciFinder® only allows you to save a limited number of hits in a single set. We therefore recommend that it only be used when stereochemistry is crucial because, in other cases, reviewing the results by hand will likely be a more efficient use of your time.

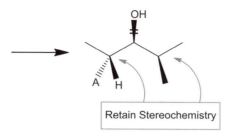

Figure 9.16   By right-clicking on an atom in Reaxys®, you get the option to invert or retain stereochemistry.

9.2.3.2.2   Stereochemistry in Reaxys® Reactions. While many reactions in Reaxys do include stereochemistry, some entries do not, so, when trying to be comprehensive, it is better to search with stereochemistry turned off and browse the resulting hits. When searching using stereochemistry, you can either draw the stereochemistry as you want it in the reactant and the product, employing atom mapping if desired, or you can use the stereochemistry option from the right-click menu. This will allow you to indicate that stereochemistry is inverted or retained in the course of a reaction and is particularly useful if you are performing a partial reaction search. For example, assume that you wish to form the structure in Figure 9.16 by forming the carbon-oxygen bond, as marked. You do not care what your starting material is, but you wish to retain stereochemistry at the two sites adjacent to the new bond. You can mark them as "stereochemistry retained", mark the bond as a bond to be formed, and you should get a desirable set of results.

9.2.4   Searching for Catalyzed Reactions

Like locating reactions containing stereochemistry, finding catalyzed reactions is complicated by the differences in organization and labeling in the various resources. As usual, it is best to consult the help files and documentation for the tool that you intend to employ, although we will attempt to present some commonalities and techniques for popular resources here. Catalysts in graphical reaction databases tend to be included in the reaction conditions, rather than being drawn out in the reaction schema, although this, too, may differ between tools. As a result, it is frequently easier to determine the most common catalyst to use for a particular type of reaction than to locate reactions catalyzed by a certain class of substance. Methodology databases and databases dealing with reagents, on the other hand, are easily searched by catalyst; you can use such resources to locate representative or types of reaction that the catalyst facilitates, or you can search by a very general reaction substructure to retrieve records for possible catalysts and reaction conditions.

9.2.4.1   Analyzing a Set of Graphical Reactions by Catalyst.   Most information retrieval systems now offer the ability to analyze results and form a histogram of frequency of occurrence of interesting aspects, and SciFinder® and Reaxys® are no exceptions. SciFinder's® default analysis for a reaction search is by catalyst, so, it is easy to see the most commonly occurring catalysts in a set of reaction results. The major downside to analyzing by catalyst in SciFinder® is that many of the catalysts are represented only by their CAS Registry Numbers, and you will need to link across to their substance records to discover their identity.

Reaxys® is much more challenging when it comes to retrieving, viewing, and analyzing reactions. Reaxys® indexes catalysts and reagents together, meaning that, when you analyze your results, you need to sort out which substances are reagents and which have been used to catalyze the reaction. Some catalysts are easy to pick out; however, many substances can be used as either catalysts or reagents, depending on the reaction in which they are employed, and it is impossible to weed out only the catalyzed reactions by a simple analysis.

9.2.4.2   Finding Reactions Catalyzed by a Specific Substance.   Because catalysts are treated differently from one resource to another, it is difficult to make generalizations about methods of locating reactions catalyzed by a particular material. We will deal with two situations, one in which you wish to locate general information about the behavior of a particular catalyst and one in which you wish to locate a specific type of reaction catalyzed by a certain substance or class of substance.

9.2.4.2.1   Locating Information on the Behavior of Catalysts. The most effective way of locating information on the behavior of a catalyst is to turn to a resource dealing with reagents or methodology. *The Encyclopedia of Reagents for Organic Synthesis* (EROS) and *Fieser's Reagents for Organic Synthesis* (Fieser) are two such tools. EROS and Fieser are both searchable by reagent name, and the electronic editions of each are keyword searchable, as well. e-EROS is easier to search online; Fieser was developed more as an e-book and is very much dependent on text, while e-EROS includes searchable structures and reactions, as well as text.

Wiley has also taken the data from EROS and repackaged it into a second series called the *Handbook of Reagents for Organic Synthesis*, available in print from the publisher and in electronic form through the Knovel Engineering & Scientific Online Reference Books. Each volume of this series focuses on a different type of transformation and presents all of the reagents from the encyclopedia that perform that type of chemistry. This can be helpful for comparison shopping purposes, although we should note that, while the articles in e-EROS are updated with new information as needed, the *Handbook* volumes remain static. It is possible to replicate the "comparison shopping" aspect of the handbook with careful keyword searching within e-EROS; for example, a keyword search for "halogenation" as a section title retrieves 45 reagents whose record has an entire section dealing with halogenation reactions. If you expand the search to all possible fields, you get even more results, although they may include some less-related possibilities.

9.2.4.2.2 Locating Graphical Reactions Catalyzed by a Specific Substance or Class of Substance. Occasionally, you want to locate only those reactions that are catalyzed by specific substances without doing a preliminary search and analyzing or refining after the fact. In essence, you wish to include the catalyst in your set of reaction conditions. This can be challenging, depending on the source that you select. In Reaxys®, you can draw the reaction into the reaction editor and then specify the catalyst that you wish to employ by using the text search field "Reagents and Catalysts" below the reaction search box. However, this can be difficult since the reagents and catalysts field requires you to input an identifier for a specific substance, and there is no very good way to search for reactions catalyzed by a particular class of catalyst.

In SciFinder®, you can search for a specific type of reaction catalyzed by a specific catalyst or class of catalyst, but you must take a roundabout path. The only way that you can draw a substance and search for it as a catalyst is to first perform a substance identifier, formula, structure, or substructure search in the substance context and then "Get Reactions" where the substance is in the role of catalyst; you are not able to set the reaction role of catalyst from within the structure editor. Once you have retrieved reactions that include the catalyst that interests you, simply refine the search with a reaction structure, including the reactant and product for a complete reaction search or either one of them if you desire to perform a partial reaction search. Example 4 demonstrates this technique.

Example 5   Using SciFinder to Locate Reactions Catalyzed by a Class of Catalyst

Assume that you are interested in finding reactions in which alcohols are converted to carbonyl functional units through reactions that employ a particular type of catalyst, shown in the scheme in Figure 9.17.

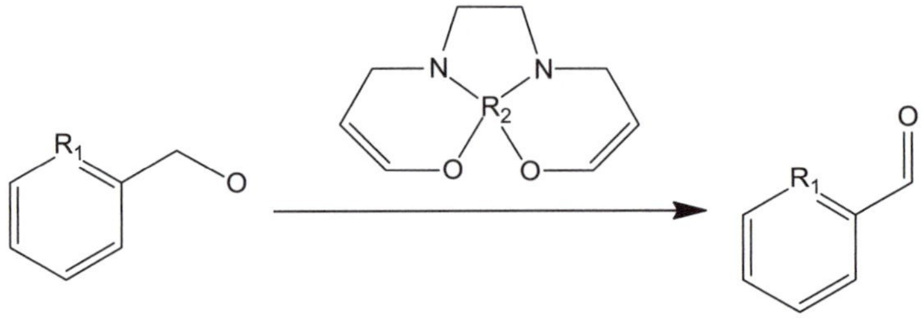

Figure 9.17   Alcohols attached to aromatic rings are converted to functional groups containing carbonyls using reactions catalyzed by the above type of catalyst. The aromatic groups may be heteroaryl, provided the heteroatom, indicated by R_1, is adjacent to the site of the reacting side-chain. The variable R_2 in the catalyst represents any first-row transition metal.

Step 1: Perform a SciFinder$^{\circledR}$ substructure search for the catalyst, following the general guidelines for searching for coordination compounds, found in Section 5.3. Refine your answer set until it contains only desirable coordination compounds, and then select the "Get Reactions" option, indicating that your structures should appear in the reaction role of catalyst.

Step 2: From your set of retrieved reactions, opt to "Refine" your hits by structure. Draw the substructure indicated in Figure 9.18, being sure to map atoms for greater precision.

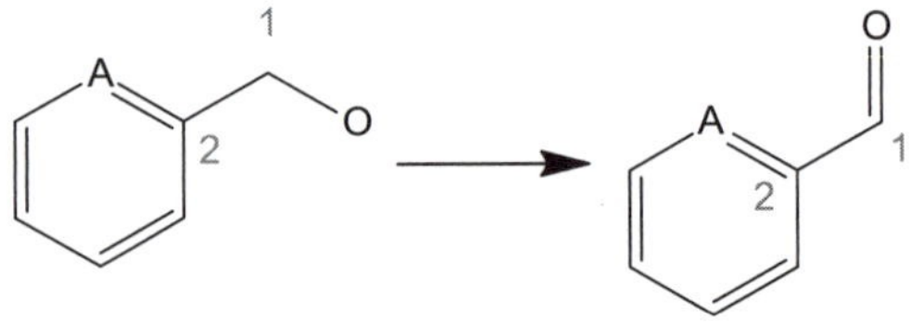

Figure 9.18   The substructures used to refine the reactions retrieved from Step 1. *A* represents any atom except hydrogen, allowing the aromatic rings to be heterocyclic, as well. Note the atom mapping from reactant to product, which will insure that the idation occurs as desired.

You should use this technique to perform a highly efficient search for reactions catalyzed by a known substance. Simply retrieve the record for the desired substance using a structure, substance identifier, or formula search of the Substances context, and then "Get Reactions" in which it is a catalyst. From your reaction results, refine by structure to retrieve the reactions you wish to catalyze. This can be more efficient than analyzing a set of reaction results by catalyst if you are only interested in reactions mediated by one particular substance, especially since SciFinder® frequently lists catalysts by CAS Registry Number.

9.3 TOOLS FOR REACTION SEARCHING

The past 20 years have been characterized by a rapid proliferation of information tools, both print and electronic, in all areas of chemistry, and the next 20 years will likely introduce still more useful books and databases. Our aim here is not to present a comprehensive bibliography of chemical reaction sources, which would be exceptionally lengthy and go out of date quickly. Instead, we will examine some of the more established and representative tools from four different types of literature: graphical reaction databases, sources of information on reagents and catalysts, methodology sources, and review literature.

9.3.1 Graphical Reaction Databases

We will define graphical reaction databases as those whose records consist of reaction schema published in the primary or tertiary literature, along with a reference to the document in which they appeared. They are generally searchable by structure and substructure, using a variety of techniques discussed in the previous section. It is important to note that the three graphical reaction databases discussed here, CASREACT, Reaxys®, and SPRESI[web], search different databases. While there is overlap in coverage, each may contain information that the others do not. Therefore, if you wish to perform a comprehensive search, it is advisable to search in multiple databases.

9.3.1.1 CASREACT. The CASREACT database, developed by the Chemical Abstracts Service, contains, at the time of this writing, more than 58 million graphical reactions, which originally appeared in the journal and patent literature, with coverage beginning in 1840 and being most comprehensive in the period from 1985 to the present.[1] These are single and multi-step reactions of organics and organometallics and include synthetic preparations of natural products, as well as biotransformations.[2] CASREACT is available through the reaction context of SciFinder®, as well as being a searchable file available from STN. Within SciFinder®, users who wish to search CASREACT may do so either by clicking the "Get Reactions" prompt from the substance or reference context or by searching the reactions context directly by means of an exact reaction search. Records include CAS Registry Numbers (usually only seen if you are searching *via* STN®) and structures for the reactants, reagents, catalysts, solvents, and products of the reaction, and may be retrieved *via* a partial or complete graphical reaction search. An increasing number of records include some experimental information; CASREACT includes this for articles published in ACS and Taylor & Francis journals, as well as patents written English, German, or Japanese from a number of authorities.[2] No matter which system you choose, you have the ability to search CASREACT *via* a simple or complex reaction structure or substructure query. In SciFinder®, you can analyze or refine your search using yield, number of steps, reaction classification, solvent, availability of experimental procedure, and bibliographic information like publication year and document type. In STN®, you have the ability to search in just about any available field by way of a complex Boolean query or by combining search sets.

9.3.1.2 Reaxys®. Reaxys'® reaction data derives from Beilstein's *Handbuch der Organischen Chemie*, Gmelin's *Handbuch der Anorganischen Chemie*, and the Patent Chemistry Database, and it is developed by Elsevier. At the time of this writing, it contains over 32 million graphical organic, organometallic, and inorganic reactions. Reaxys® reaction records include structures and identifiers for reactants and products, with reagents, catalysts, and solvents included as part of the reaction conditions. The system is extremely robust, and it is possible to specify conditions such as reagents and catalysts, solvents, temperature, pressure, time, and pH value, as well as yield, reaction classification, and number of steps and stages at the time of search, although most chemists simply draw the graphical reaction scheme with no refinements. If you choose to do this, you can refine after the fact by structure or substructure, yield, reagent or catalyst, solvent, reaction type, number of steps, commercial availability of reactant and product, and various bibliographic fields for the document that contains the reaction. You should bear in mind that not every reaction has information in all of the searchable fields, so, including reaction conditions in your search may unduly limit the scope of your results. If you wish to do a comprehensive search for a type of reaction, you should usually leave out the conditions. If, however, you are interested in locating reactions run in very specific conditions and are not concerned with being comprehensive, you should try to input a specific search from the start.

Reaxys® also includes a text field called "Subject Studied", which can be extremely interesting. This indicates the focus of the paper with respect to the reaction retrieved and includes such values as product distribution, kinetics, and mechanism. Since not all reactions include subjects studied, you should avoid this if you are performing a comprehensive reaction search; however, it is extremely helpful when looking for mechanistic or kinetic studies, which can be difficult to retrieve without a combination of reaction searching and text searching in the abstracts of the records retrieved.

9.3.1.3 SPRESI^{web}. SPRESI^{web}, from the German company Infochem, is a much smaller database than either of the previous two, comprising only about 11.8 million substances and 4.2 million reactions from about 1,300 different literature sources at the time of this writing. Its name is an acronym for "**Sp**eicherung und **R**echerche **S**trukturchemischer **I**nformation", which translates to "Storage and Retrieval of Chemical Structure Information".[3] The reaction search interface includes the ability to use structure and substructure techniques to perform either partial or complete reaction searches. Once one has a set of reaction search results, it is possible to classify them according to atoms in the immediate neighborhood of the reaction centers. The "broad" category includes only the reaction centers, the "medium" category extends one atom from the reaction centers, and the "narrow" category extends out two atoms from the reaction centers. This allows a user to get a sense of the types of transformations represented in their hits.[4] SPRESI^{web} also allows users to locate transformations similar to a reaction of interest and to find name reactions.[5] Infochem has also created a very nice mobile application that allows all users to access a certain amount of the database free of charge.

9.3.2 Sources of Information on Reagents and Catalysts

At times, you may have a general idea of how you wish to proceed with a transformation, but you would like to locate a good reagent or system of reagents that will perform the reaction for you. You can choose to look up the information in a graphical reaction database, but it can be more efficient to search for a general reaction class or substructure using a source of reagent information. These sources have the advantage of giving you fairly comprehensive information about the reagent and its reactivity, which can help you when selecting one for a novel reaction. The two largest reagent sources are the *Encyclopedia of Reagents for Organic Synthesis*[6] and *Fieser's Reagents for Organic Synthesis.*

9.3.2.1 The Encyclopedia of Reagents for Organic Synthesis (EROS). The *Encyclopedia of Reagents for Organic Synthesis*, or EROS, published by Wiley and edited by Leo Paquette, is now in its second edition and is available in both print and electronic format. It has about 4,000 detailed articles about a variety of reagents and catalysts used in the synthesis of organic molecules.[6] Each record begins with an "abstract", which generally includes the structure, CAS Registry Number, formula, molecular weight, InChI, and InChIKey; a brief description of the function of the reagent or catalyst; physical data and solubility; the form in which it is available or preparatory information or references; purification information; and some basic safety information, including handling and storage guidelines. The rest of the record details the types of reactions that it facilitates and includes graphical reactions, tables, and descriptive text. All told, the database contains some 70,000 examples of reactions.[7]

Both the print and electronic formats of EROS are arranged alphabetically by reagent name, and you can browse the articles easily by name in both. The print includes a reagent formula and a subject index. e-EROS allows you to search by keyword in the text of the articles, a context that you can also use to search for CAS Registry Numbers, InChIs, and InChI Keys. The structure search context permits complete structure or substructure searches for substances and reactions, and its text search fields beneath the structure editor enable you to enhance your search with some reaction conditions, including yields, catalysts, solvents, and temperatures from the reactions.

EROS and e-EROS are best used when you have an idea of which reagent you wish to employ for a particular reaction and you want to know more about its reactivity. The fact that the physical properties, purification, and major safety hazards are all present in the same place as the examples of reactions makes this a good first-stop for information about potential reagents. The information published in EROS has also been reproduced in a series called the *Handbooks of Reagents for Organic Synthesis*, which aim to collect into a single volume all reagents that perform a certain type of reaction. While this can help with comparison shopping, a simple text search in e-EROS will perform a similar function.

9.3.2.2 Fieser's Reagents for Organic Synthesis. Fieser's *Reagents for Organic Synthesis* is an older work than EROS; it was begun in

1967 by Louis F. Fieser and Mary Fieser, and some people refer to it as *Fieser and Fieser*. Unlike EROS, which forms one complete encyclopedia, *Fieser* has been published as a book series, with new volumes being released periodically. To make it easier to locate information on the various reagents published in the volumes, *Fieser* includes collective indices to groups of volumes, thus minimizing the number of books in which you need to search to see if information on your chosen reagent has been updated. *Fieser* is now available online through the Wiley Major Reference Works, either as a subscription database or as a one-time purchase with update fees whenever new volumes are released. *Fieser*'s online interface is much more clunky and restrictive than that of e-EROS, and it does not permit structure searching. However, it is still a valuable source of information on a wide variety of reagents, and it should not be ignored.

9.3.3   Methodology Sources and Review Literature

The review literature, including sources that describe methodology, is a wonderful place to begin exploring a type of reaction that interests you. As has already been described, a reader of the review literature puts himself or herself into the hands of the author or editor of the work in question and, in return, gets a concise review of the primary literature written to date. A reader of the review literature must therefore rely on the fact that the source consulted is comprehensive and presents a balanced view of the important primary works in that subject.

Review literature is particularly useful in the area of chemical reactions; because of its cumulative and judgmental nature, it is frequently used to compare different methods of performing a transformation of interest. Therefore, it is one of the few types of literature to which you can turn to learn about things that do not work particularly well, and it can be very helpful in determining the best conditions under which to perform a particular type of reaction. The largest set, by far, is *Houben Weyl's Methoden der Organischen Chemie*, whose latest segment, *Science of Synthesis*, forms a helpful stand-alone review source; however, there are many other comprehensive sets that can assist the synthetic chemist, several of which are presented in the last section here. We have decided to deal only with "formally" published work in this chapter, although social media is beginning to gain momentum as review literature, and many synthetic organic chemists monitor the blog "Totally Synthetic" (http://totallysynthetic.com/blog/), whose author and readers comment on recent advances in total synthesis.

9.3.3.1 Houben Weyl's Methoden der Organischen Chemie and Science of Synthesis. Houben Weyl *Methoden der Organischen Chemie*, published by Thieme, is one of the largest review sources in existence. The first edition was published in 1909 by Theodor Weyl, and it contained one volume dealing with analytical methods and purification, and two volumes covering oxidation and reduction reactions, oxygen, sulfur, halogen, and nitrogen compounds, and organometallics. In 1913, Heinrich Houben began the second, four-volume edition on the same topics, the first volume of which was released in 1921, and, in 1924, the first of four volumes of the third edition was released. The fourth edition, published between 1958 and 1986, considerably expands the first three editions, having a total of 67 volumes, and, shortly after its completion, the series grew to include the "E-series", which were several groups of volumes devoted to the synthesis of particular types of substance. The most recent addition to the Houben Weyl series is *Science of Synthesis: Houben Weyl Methods of Molecular Transformations*, which began publication in 2000, and which continues to be updated today.[8] We will refer to this part of the series simply as "*Science of Synthesis*" and will call the first four editions and the E-series "*Houben Weyl*". Prior to 1990, *Houben Weyl* was published only in German, but in 1990, Thieme began publishing in English as well. *Science of Synthesis* is published in English.

Houben Weyl is organized according to the functional group that you wish to synthesize. Thieme released a printed users' guide in 2001, which explains exactly how the functional groups are arranged in the printed volumes,[9] and, in fact, some of the volumes actually include structural diagrams of key units on their spines. *Houben Weyl* is now available online through Thieme's *Science of Synthesis* online product, and it has a detailed browsable and searchable table of contents in English, which helps non German-speakers navigate the organization and get to the information of interest to them.

Like the rest of *Houben Weyl*, *Science of Synthesis* is arranged according to the group that is to be constructed. This has the effect of grouping together all methods for making a particular type of product, which makes it quite easy to use, given the fact that most people turning to this type of literature are exploring various methods of synthesizing something. *Science of Synthesis* was released simultaneously in print and electronic format, a volume at a time, from 2000 to 2008. The electronic edition is searchable by structure and text; when searching by structure, you may enter an exact or substructure substance or reaction search, and text fields may be used either to refine the structure search or in lieu of it. At the time of this writing, the text search capabilities include searching by CAS Registry Number, catalyst, solvent, temperature, and reaction yield; searching by full text or name reaction; and searching for a specific reference within the series using the volume, author, and page number.

9.3.3.2 Organic Syntheses and Organic Reactions. Both *Organic Syntheses* and *Organic Reactions* are multivolume series in organic chemistry begun by the same chemist, Roger Adams, that aim to present a broad scope of organic reactions from the primary literature in a concise, easy-to-use manner. *Organic Syntheses* is an annual series that originated in 1921 and is focused on synthetic methods that work well. Each synthetic method selected is presented clearly enough for a scientist at the graduate-student level or higher to be able to successfully follow the procedure as written. Methods chosen for inclusion are tested in the laboratory of a member of the editorial staff, so every method has been shown to work in at least two laboratories, those of the submitting scientist and the "checker".[10] In addition to the annual volumes, *Organic Syntheses* publishes a set of collective volumes, called *Organic Syntheses Collective*. Originally, one was published every ten years, but, in 1984, the frequency increased to one every five years. The collective volumes bring together information originally presented in the annual volumes, but they also include improvements upon the methods presented. It is, therefore, essential that you do not confuse references to the annual and collective volumes when trying to locate a cited preparation. *Organic Syntheses* is a non-profit "Membership Corporation" and, as such, distributes the series at very low cost, making it available to the ACS Division of Organic Chemistry for distribution to its members. In 1998, they began making all annual and collective volume contents available to the world free of charge, at http://www.orgsyn.org.[10] You can search by bibliographic information, structure or substructure, graphical reaction, type of reaction, formula, chemical name, CAS Registry Number, or keywords in the text. Researchers who wish to have more powerful search capabilities or who want to simultaneously search *Organic Syntheses* volumes along with other Wiley content can also license the content from Wiley.

Organic Reactions, on the other hand, uses each volume to present "collections of chapters each devoted to a single reaction, or a definite phase of a reaction, of wide applicability".[11] They present the information from a synthetic perspective, but, unlike *Organic Syntheses*, the procedures included are not rigorously tested. Instead, an author knowledgeable in the subject presents the benefits and limitations of each technique, as well as important variations on the method.[11] However, probably the most important aspect of this resource is the fact

that the author of each chapter has performed a comprehensive literature search for articles that employ the reagent and has presented their
references in tabular form. The print series employs a detailed table of
contents and index to help searchers locate the reaction of interest. In
addition to the print volumes, which are released periodically, Wiley has
begun offering *Organic Reactions* as an online product searchable by
structure, substructure, reaction CAS Registry Number, chemical name,
reaction type, and reaction conditions.[12]

9.3.3.3 Other Comprehensive Sets. Starting in the late 1970s and
continuing to the present time, a variety of publishers have put out nice
treatises dealing with various segments of the chemical literature. These
"Comprehensive" sets provide an excellent starting point when researching a topic, and many of them deal with organic and organometallic
reactions. However, even those devoted to a particular type of substance
frequently present information about the synthesis and reactions of
those substances. The following list introduces some of these reference
works, which include information on chemical reactions, although some
devote more time and attention to the topic than others. The works
dealing specifically with synthesis and reactions are presented first.

9.3.3.3.1 Comprehensive Organic Works
Comprehensive Organic Chemistry[13]

This multi-volume set was published in 1979, and it is the most general
of the comprehensive works dealing with organic reactions. The editors
state their goal as being to help chemists to stay abreast of the growing
body of literature in synthetic chemistry. The work reviewed is entirely
experimental; they omitted to cover any theoretical organic chemistry
because that area changes too rapidly for the production of a work of
this nature. *Comprehensive Organic Chemistry* includes a reaction and a
reagent index, and the indices themselves contain additional references
to the primary literature, which do not appear in the text of the volumes.

Comprehensive Organic Synthesis[14]

This work was created to build upon the information presented in
Comprehensive Organic Chemistry. The editors declare, in the preface,
that the object of this book is "to focus on transformations in the way
that synthetic chemists think about their problems". It is roughly
organized according to carbon-carbon bond formed, heteroatom
introduced, or heteroatom exchange, although, depending on the nature
of the transformation being described, the classification of a particular
synthesis can seem somewhat arbitrary.

Comprehensive Organic Functional Group Transformations I[15] and II[16]

This work was published in 1995 and contains seven volumes devoted to the reactions involving various functional groups. It is organized "on the basis of formation or rupture of bonds to a carbon atom".[15] The organizational system also takes into account the number and identity of heteroatoms in the group, the degree to which carbon is coordinated, and the "Latest Placement Principle". Each set has a detailed table of contents, as well as a subject index.

Larock's Comprehensive Functional Group Transformations[17]

Richard Larock's classic treatise on functional group transformations, now in its second edition, is organized roughly by transformation. Larock selects reactions that either have a general scope or are extremely unique and show great promise for synthetic applications. All reactions included give yields of at least 50% and employ reagents that are either commercially available or easy to prepare and use.[17] Larock purposefully omits protecting groups and heterocyclic chemistry from his work, stating that many useful review sources already exist in these areas.

Greene's Protective Groups in Organic Synthesis[18a]

The first edition of *Protective Groups in Organic Synthesis* was written in 1981 by Theodora Greene. Greene went on to collaborate with Peter G. M. Wutts on the next two editions of the book[18b] (orgsyn.org/obits/greene.pdf), and Wutts published the current, fourth edition two years after her death. It is organized by functional group to be protected, and each entry includes information about both protecting and deprotecting the group.

Protecting Groups[19]

Philip Kocienski's book on protecting groups is the other classic text in the area of protecting groups. It is currently in its third edition, and, like Greene's book, it is organized by functional group.

9.3.3.3.2 Comprehensive Works Dealing with Organometallics, Coordination Compounds, and Catalysts

Comprehensive Organometallic Chemistry I[20], II[21], III[22]

The goal of these series of volumes, published in 1982, 1995, and 2006, respectively, is to present a comprehensive view of the organometallic literature. As such, they present a good deal of information about the synthesis of the ligands and complexes themselves; however, since organometallic substances are frequently used to facilitate organic reactions, information about their use as catalysts is also given great prominence.

Comprehensive Coordination Chemistry I[23], II[24]

These works deal with the chemistry of coordination compounds. Within the first set, volume one includes information on reaction mechanisms, volume two has information on ligand synthesis and coordination, and volume six includes a chapter entitled "Uses in synthesis and catalysis." In addition to a cumulative subject index, there is a cumulative formula index, making it easy to find information on each substance included. The second set is subtitled, "From Biology to Nanotechnology," and it purports to include "new techniques of synthesis and characterization".[24]

Comprehensive Asymmetric Catalysis[25]

This is a set of three volumes with two supplements edited by Eric Jacobsen, Andreas Pfaltx, and Hiyashi Yamamoto. It is organized according to the function that the catalyst performs, examples of which include hydrogenation of carbon-carbon double bonds, Aldol reactions, and epoxidations. It presents various catalyst systems and discusses how each behaves under particular conditions. The set also comes with an accompanying disc set.

Compendium of Chiral Auxiliary Applications[26]

This set of volumes scours the synthetic literature up to 2000 looking for applications of as many chiral auxiliaries as possible. The author attempts to present a comprehensive view of the literature, and, as such, he does not make judgments about the level of selectivity that an auxiliary provides. It is organized by type of reaction, with the last chapter used as a catch-all for reactions that did not logically fall into any of the previous categories. Each record includes information about the substrate, the reagents and conditions employed, and the results of the reaction, including stereoselectivity, yield, and stereocenter configuration. The work references 2,700 primary articles.[26]

9.3.3.4 Sources Dealing with Name Reactions. In addition to the ability to use name reactions to search through many of the tools listed in the previous sections, there are a number of excellent books devoted exclusively to the mechanisms and applications of various name reactions. A search of your library catalog using the phrase "name reaction" will likely retrieve many options. Some of the classics include: Mundy's *Name Reactions and Reagents in Organic*

Synthesis,[27] now in its second edition; Jie Jack Li's *Name Reactions: A Collection of Detailed Reaction Mechanisms*,[28] the fourth edition of which has been expanded to include synthetic applications; and Alfred Hassner's *Organic Syntheses Based on Name Reactions*,[29] the first two editions of which were published as part of the *Tetrahedron Organic Chemistry Series*. However, there are also some excellent relative newcomers on the scene. László Kürti and Barbara Czakó's *Strategic Applications of Named Reactions in Organic Synthesis*[30] presents the original papers in which the reactions appear, as well as important applications in later syntheses. Wiley has also published several edited books and a book series, the Wiley Series on Comprehensive Name Reactions, dealing with name reactions. Most recently, they published a three-volume set by Zerong Wang called *Comprehensive Organic Name Reactions and Reagents*,[31] which discusses 701 name reactions and claims to be the most up-to-date work of its kind.

9.4 CONCLUSION

The graphical reaction search techniques mentioned in the second section of this chapter are tricks developed over time. They should be used to get you started searching for reaction information in graphical databases, but, as you become a more experienced searcher, you may discover some other methods of retrieving useful reaction information. It is important to remember that, the more general a search is, the more information you will recall; and the more exact your specifications, the less you will retrieve, although those things that do turn up will be extremely precise. As with any other type of search, you will need to balance recall with precision, and the balance may shift depending on the reason for your search. If you are searching for a preparation that will take you to the next step of your synthesis, you may want to be very specific in your search, and you can stop once you have found a reaction that fits your needs. However, if you plan to announce in a funding proposal that nobody has ever performed a similar reaction as efficiently as you have, you will need to be very exhaustive in your searching, selecting more general substructures and being more creative in your thinking about the substrate and product.

The information landscape is changing rapidly over time, and graphical reaction search tools continue to evolve. Many companies are beginning to incorporate tools into their products to help researchers simplify their workflow. For example, both the Chemical Abstracts Service and Elsevier's Reaxys® group have implemented synthesis planners, through which a scientist can generate an attractively organized synthetic scheme simultaneously with searching the literature. Likewise, many companies are developing search algorithms that allow a searcher to start with a desired reaction and locate "more like this", specifying greater and lesser degrees of similarity (retain only the reacting center; retain the reacting center and one atom out, *etc.*) and to analyze your results, grouping reactions by criteria that you specify. It is imperative that you stay up-to-date with new tools and trends; despite the fact that your previous workflows may still function, new developments may allow you to develop even more effective and time-saving strategies.

As was mentioned in the introduction to the chapter, the tools mentioned are examples of several varieties of reaction sources you may wish to consult, but they by no means constitute an exhaustive bibliography on the subject. A number of "comprehensive" sets on various topics, such as *Comprehensive Heterocyclic Chemistry*, *The Porphyrin Handbook*, and Patai's *Chemistry of Functional Groups* series, include information on the synthesis of various classes of molecule, as do many monographs and smaller, two- and three-volume sets. Some of these works are indexed in databases like *Chemical Abstracts*; others can be found using a text search of your library catalog. The Royal Society of Chemistry produces two current awareness tools, *Catalysts & Catalyzed Reactions* and *Methods in Organic Synthesis*, which present recent publications in an organized way and are terrific for people who like to stay up-to-date by browsing. Finally, remember that your librarian or information professional can also help direct you to resources held locally that will assist you with reaction searching.

ACKNOWLEDGEMENTS

The author would like to thank Grace Baysinger for her input and assistance in preparing this manuscript.

REFERENCES

1. See the database summary sheet, at http://www.cas.org/File%20Library/Training/STN/DBSS/casreact.pdf for more information.
2. Get More Content in SciFinder: Reactions, http://www.cas.org/products/scifinder/content-details#reaction, Accessed September 25, 2012.
3. Spresi FAQ, http://www.spresi.com/, Accessed September 15, 2012.
4. ICClassify: The InfoChem reaction classification program, http://infochem.de/content/downloads/classify.pdf, Accessed October 11, 2012.

5. Structure Input and Search Documentation, http://infochem. de/content/downloads/icfsetutorial.pdf, Accessed September 15, 2012.

6. L. A. Paquette, D. Crich, P. L. Fuchs and G. A. Molander, eds., *Encyclopedia of Reagents for Organic Synthesis*, John Wiley & Sons, Chichester, 2009.

7. e-EROS: Encyclopedia of Reagents for Organic Synthesis, http:// onlinelibrary.wiley.com/book/10.1002/047084289X, Accessed September 26, 2012.

8. *Science of Synthesis: Houben-Weyl Methods of Molecular Transformations Guidebook*, Georg Thieme Verlag, Stuttgart, 2002.

9. *Houben-Weyl Methods of Organic Chemistry Users' Guide*, Georg Thieme Verlag, Stuttgart, 2001.

10. R. L. Shriner, R. H. Shriner and J. P. Freeman, History of Organic Syntheses, http://www.orgsyn.org/about.asp, Accessed September 25, 2012.

11. R. Adams, Introduction to the Series, http://onlinelibrary.wiley. com/book/10.1002/0471264180/homepage/ORIntrototheSeries.pdf, Accessed September 26, 2012.

12. About *Organic Reactions*, http://onlinelibrary.wiley.com/book/ 10.1002/0471264180, Accessed September 25, 2012.

13. D. Barton and W. D. Ollis, eds., *Comprehensive Organic Chemistry: The Synthesis and Reactions of Organic Compounds*, Pergamon Press, Oxford, 1979.

14. B. M. Trost and I. Flemming, eds., *Comprehensive Organic Synthesis: Selectivity, Strategy & Efficiency in Modern Organic Chemistry*, Pergamon, Oxford.

15. A. R. Katritzky, O. Meth-Cohn and C. W. Rees, eds., *Comprehensive Organic Functional Group Transformations I*, Pergamon, Cambridge, 1995.

16. A. R. Katritzky and R. A. J. Taylor, eds., *Comprehensive Organic Functional Group Transformations II*, Elsevier, Oxford, 2005.

17. R. C. Larock, *Comprehensive Functional Group Transformations: A Guide to Functional Group Preparations*, John Wiley & Sons, New York, 1999.

18. (a) P. G. M. Wutts and T. W. Greene, *Greene's Protective Groups in Organic Synthesis*, John Wiley & Sons, Hoboken, NJ, 2007; (b) F. D. Greene, *Theodora Watmough Greene*, http://orgsyn. org/obits/greene.pdf, Accessed August 23, 2013.

19. P. J. Kocienski, *Protecting Groups*, Georg Thieme Verlag, Stuttgart, 2005.

20. G. Wilkinson, F. G. A. Stone and E. W. Abel, eds., *Comprehensive Organometallic Chemistry I: The Synthesis and Structures of Organometallic Compounds*, Pergamon Press, Oxford, 1982.
21. E. W. Abel, F. G. A. Stone and G. Wilkinson, eds., *Comprehensive Organometallic Chemistry II: A Review of the Literature 1982–1994*, Pergamon Press, Oxford, 1995.
22. G. Wilkinson, F. G. A. Stone and E. W. Abel, eds., *Comprehensive Organometallic Chemistry III: The Synthesis and Structures of Organometallic Compounds*, Pergamon Press, Oxford, 2006.
23. G. Wilkinson, R. D. Gillard and J. A. McCleverty, eds., *Comprehensive Coordination Chemistry: The Synthesis, Reactions, Properties and Applications of Coordination Compounds*, Pergamon Press, Oxford, 1987.
24. J. A. McCleverty and T. J. Meyer, eds., *Comprehensive Coordination Chemistry: From Biology to Nanotechnology*, Elsevier, Oxford, 2004.
25. E. N. Jacobsen, A. Pfaltz and H. Yamamoto, eds., *Comprehensive Asymmetric Catalysis*, Springer Verlag, New York, Berlin, Heidelberg, 1999.
26. G. Roos, *Compendium of Chiral Auxiliary Applications*, Academic Press, San Diego, 2002.
27. B. P. Mundy, *Name Reactions and Reagents in Organic Synthesis*, Wiley Interscience, Hoboken, NJ, 2005.
28. J. J. Li, *Name Reactions: A Collection of Detailed Mechanisms and Synthetic Applications* Springer-Verlag, Berlin, 2009.
29. A. Hassner, *Organic Syntheses Based on Name Reactions: A Practical Guide to 750 Transformations*, Elsevier, Oxford, Boston, 2012.
30. L. Kürti and B. Czakó, *Strategic Applications of Named Reactions in Organic Synthesis: Background and Detailed Mechanisms*, Elsevier Academic Press, Amsterdam; New York 2005.
31. Z. Wang, *Comprehensive Organic Name Reactions and Reagents*, John Wiley & Sons, Hoboken, NJ, 2009.

CHAPTER 10

A Practical Primer to BLAST Sequence Similarity Searching

DIANE C. REIN

University at Buffalo, Health Sciences Library, Abbott Hall, 3435 Main St., University at Buffalo, Buffalo, New York, 14214, US
Email: drein@buffalo.edu

10.1   INTRODUCTION

Unlike literature (bibliographic) records, which contain only text, bioinformatics records contain both data (the sequence) and text that describes that sequence. Separate mechanisms have been developed to search either the text (in a fashion very similar to literature databases), or for the data (sequence) itself.

Literature database systems are built upon the fact that each component of the record is immutable. The title of the article, where, when, and by whom it was published, is not going to change. Stating the obvious, each journal article record in a literature database is unique. Search engines are built to match text in a query to identical text in a literature record. In most sciences, molecular uniqueness is also true. In physics, a muon is a muon. In chemistry a carbon atom is a carbon atom. Search algorithms in these disciplines are built upon this concept whether they are text-based or structural-based. Biological systems, however, are completely different as they are built upon enhancing diversity. No two organisms are exactly identical, including their

Chemical Information for Chemists: A Primer
Edited by Judith N. Currano and Dana L. Roth
© The Royal Society of Chemistry 2014
Published by the Royal Society of Chemistry, www.rsc.org

genome and gene sequences. Even within the same organism, genes and proteins can have variable sequences of the same protein in different cellular, tissue, or organ compartments. In the biological sciences, if one were to use a DNA or protein sequence as a query and use a search engine based upon identity as literature databases do, the only sequence retrieved would be the query sequence itself. Thus, in bioinformatics, searching sequence data in bioinformatics records are based upon sequence *similarity* searching.

Typically, the product from the laboratory research bench is a DNA or protein sequence that complements some type of biological function or assay. Often, the researcher has little, if any, knowledge about the sequence beyond the actual sequence itself. Sequence similarity programs first retrieve similar sequences from a database and then align them against each other. Wherever the query sequence aligns with a database sequence of known structure, sequence or biological activity, the query sequence can be predicted to possess the same properties as the retrieved database sequence. Thus, one of the major uses of sequence similarity searching is to locate similar sequences for which more knowledge is known. Uses for sequence similarity searching include, but are not limited to:

- Assigning function to unknown sequences, including:
  - locating mutations and/or genetic variations;
  - identifying similar sequences from different organisms or tissues; and
  - identifying the various protein domains of a sequence.
- Comparing two or more sequences to each other.
- For those working on protein modeling or in protein engineering, locating 3D structure records for any given sequence.
- Locating similar sequences that bind drugs, organic substances, ions, or nucleic acids. This again aids in assigning function or finding the best prototypic sequence to initiate studies on, for example, docking of drugs to target proteins.
- Identifying that portion of a sequence which would best support cloning through the use of the polymerase chain reaction (PCR).
- Cleaning suspected cloning vector sequences from cloned sequences.

The Basic Local Alignment Sequence Tool (BLAST), first developed in 1990[1] by the National Center for Biotechnology Information (NCBI) and subsequently algorithmically restructured in 1997,[2] is the most commonly used sequence similarity tool in use today. Found at NCBI as a suite of BLAST tools, BLAST now populates bioinformatics data-bases worldwide as the sequence similarity search tool of choice. Accessed over the Internet, BLAST is preconfigured with a set of default parameters to guarantee functionality for a wide general audience. Most researchers run BLAST with the default settings, either never realizing the wide array of available options to adjust and tailor the algorithm to individual needs, or unsure how to change what setting for what reason. As important as running a BLAST search is, it is only the first step in assigning a biological function to a biosequence. Following a BLAST search, researchers must also develop effective and efficient strategies to migrate laterally through the extensively hyperlinked NCBI databases to locate the considerable knowledge present at NCBI for assigning function to, and/or gathering new knowledge about, the sequence under investigation.

This chapter begins with sections designed to enable a general understanding of sequence similarity searching, followed by a detailed explanation of how BLAST works. The aim is to fill the existing gap between BLAST theory and practice that currently exists for many users.

The last half of the chapter is presented as a hands-on self-training BLAST tutorial designed to be read while performing the BLAST exercises at a computer. Although the focus is to NCBI protein BLAST (blastp), the techniques learned can be applied to all NCBI BLAST interfaces, including nucleotide BLAST. With few exceptions, they can also be applied, albeit through different interfaces and mechanisms, at most BLAST resources around the world. The tutorial incorporates strategies to empirically determine effective analyses of BLAST results, as well as techniques to assign biological function(s) to sequences of choice retrieved through a BLAST search.

10.2 BEFORE SEARCHING: DISTINGUISHING BETWEEN IDENTICAL, SIMILAR (IDENTITY), AND HOMOLOGOUS SEQUENCES

Often used interchangeably by researchers, the concepts of "identity", "similar" and "homologous" sequences are distinct. BLAST search results report degrees of similarity and provide mechanisms to locate homologous sequences. Distinguishing between these terms is critical to properly evaluate BLAST results, as well as to understand the funda-mental mechanics underpinning the BLAST mathematical expression and its statistical operations.

10.2.1 Defining Identical

Identical is the degree to which sequences are exactly the same (invariant). Two sequences that are identical would have exactly the same nucleotide or amino acid sequence with no substitutions. Except for very short sequences (approximately nine amino acids or lower), the chance of two or more biological sequences being identical (exact match) along their length is extremely low. Although they can have small regions of exact matches, they can be anticipated to be comprised predominantly of stretches of similar or mismatched sequences. As a result, the terms "identical" or "exact match" are seldom used, as they rarely exist.

10.2.2 Defining Similarity

Similarity is the degree to which two or more biological sequences are related to each other. It involves no concept of historical (evolutionary) relationships among sequences. Similarity permits conservative sequence substitutions. For proteins, this usually means substituting a neutral amino acid for another neutral amino acid, or a polar amino acid for another polar amino acid. *The percent to which sequences are similar is often reported within BLAST searches as "% identity", which is not to be confused with "identical" as described above.*

Sequence similarity implies functional similarity, but does not guarantee it. When analyzing sequence similarity search results from any sequence similarity algorithm, including BLAST, it is important to remember that:

1. Similar sequences can have a different function in a different tissue or organism.
2. Different (dissimilar) sequences can have the same function in a different tissue or organism.
3. Different sequences can have the same 3D structure.

Similarity is what is observed in a dataset at the time the search is run. Biosequence databases are quite dynamic, with constantly changing content. BLAST searches run at later dates, will be against a different dataset of records. Different similarities can be expected to be found, resulting in different results. Rerunning BLAST searches to continually reassess sequence function is an important strategy, as new searches will yield new sequences as well as new knowledge for existing sequences.

10.2.3 Defining Homology

Homology invokes the concept of the historical (evolutionary) biological relationships between sequences. This can inform on assigning a biological function to sequences under investigation. *Phylogeny* is the description of those biological relationships over evolutionary time. Phylogenetic relationships are typically derived by software programs that cluster sequences into related groups, which are then visualized as a "tree". NCBI provides an opportunity to phylogenetically analyze BLAST search results through an option called the *distance tree of results*. To fully assess homology following a NCBI BLAST search, NCBI provides researchers the ability to migrate laterally from single hits retrieved by BLAST to a corresponding sequence record NCBI HomoloGene resource.

10.3 THE MAJOR SEQUENCE SIMILARITY ALGORITHMS AND THEIR (CONTINUING) RELATIONSHIP WITH BLAST

All sequence similarity algorithms share the common denominators of first locating similar sequences within a database, retrieving them, and then aligning them to each other such that areas of similarity are easily visualized within the results. The sequence used to search is called the *query sequence*. The sequences retrieved from the database are called the *subject sequences*. Sequence similarity algorithms that align along the entire length of the query and subject sequences are called *global alignment algorithms (programs)*, while those that seek only short areas of similarity are called *local alignment algorithms (programs)*.

NCBI BLAST is at the end of a lineage of sequence similarity algorithms (Figure 10.1) which began with the development of the Needleman-Wunsch sequence similarity algorithm in 1970.[3] As BLAST gained popularity due to its rapid ability to return results, particularly from very large sequence databases, and began to populate bioinformatics databases worldwide, the use of the earlier sequence similarity algorithms declined. Many researchers do not realize that they are still available for use, can play a strategic role in confirming BLAST results, verifying sequence alignments, supplementing BLAST searches, or as viable alternatives when BLAST search results are less than satisfactory.

10.3.1 The Needleman-Wunsch Algorithm

Comparing two sequences to each other, either nucleotide or protein, Needleman-Wunsch aligns sequences along their entire respective lengths, seeking the best possible alignment. Thus, Needleman-Wunsch is a global alignment algorithm. Its limitations are that it can only compare a query sequence to another sequence, one at a time. It is also computationally intense, particularly with longer sequences. Between these two caveats, it cannot support research in "real-time" nor retrieve

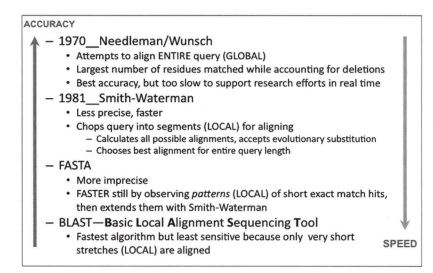

ACCURACY

- 1970__Needleman/Wunsch
  - Attempts to align ENTIRE query (GLOBAL)
  - Largest number of residues matched while accounting for deletions
  - Best accuracy, but too slow to support research efforts in real time
- 1981__Smith-Waterman
  - Less precise, faster
  - Chops query into segments (LOCAL) for aligning
    - Calculates all possible alignments, accepts evolutionary substitution
    - Chooses best alignment for entire query length
- FASTA
  - More imprecise
  - FASTER still by observing *patterns* (LOCAL) of short exact match hits, then extends them with Smith-Waterman
- BLAST—Basic Local Alignment Sequencing Tool
  - Fastest algorithm but least sensitive because only very short stretches (LOCAL) are aligned   SPEED

Figure 10.1   A short history of sequence searching. Developed in 1970, Needleman-Wunsch, a global alignment tool and still considered the most accurate algorithm, is the most computationally intense but unable to support laboratory bench research in real time. Subsequent generations of algorithm development introduced various renditions of local alignment tools, including Smith-Waterman and FASTA, resulting in increased computation rates, but at the sacrifice of accuracy. BLAST readily supports both real-time research and can rapidly search the large sequence databases currently in existence.

multiple records from the large datasets in existence today as typified by GenBank. However, many consider it to yield the best possible alignment between two sequences. It has always been available at the European Bioinformatics Institute (EBI, http://www.ebi.ac.uk/Tools/sss/) and is now available in a "tweaked" version that reduces computational memory usage from the NCBI BLAST home page under the Specialized BLAST section (http://blast.ncbi.nlm.nih.gov/).

10.3.2   The Smith-Waterman Algorithm

The Smith-Waterman algorithm[4] is a variation on Needleman-Wunsch. Instead of determining similarity along the entire length of a sequence, it searches for regions of local similarity between the Query sequence and a Subject sequence, calculates all areas of local similarity, and then chooses the best alignment for the entire query sequence length. Similarity algorithms that align sequences can identify those local areas of a sequence that remain conserved over evolution (homologous) when the rest of the sequence has diverged and is dissimilar.

Smith-Waterman was a major innovation in sequence similarity searching as it extended similarity searching into locating homologous sequences in organisms across an evolutionary distance, greatly enhancing assigning functions to unknown sequences. Unlike Needleman-Wunsch, Smith-Waterman is capable of searching against large sequence databases. It is faster than Needleman-Wunsch but sacrifices accuracy for speed. Created in 1981 when sequence databases were very small, it remains too computationally intense to regularly search the large sequence databases in existence today. Current versions of Smith-Waterman, computationally adjusted for speed increases to run against large datasets, are available as SSEARCH at EBI (http://www.ebi.ac.uk/Tools/sss/).

10.3.3 The FASTA Algorithm

Faster than Smith-Waterman but less sensitive, the FASTA algorithm was developed to permit similarity sequence searches in large databases in support of "real-time" researchers and their use of microcomputers at the research bench.[5,6] Although not used as extensively as it was prior to the creation of BLAST, most researchers using sequence similarity searching will be familiar with one of FASTA's lasting legacies: The query input format known as the FASTA format which has become the standard query input for most sequence similarity algorithms in use today, including BLAST. FASTA is a *local sequence similarity algorithm* that observes *patterns* of local similarity hits, aligning only those with highest scoring values. FASTA is considerably more sensitive than BLAST in finding similar sequences as it tends to locate longer stretches of local areas of similarity with higher *percent identities*.

FASTA assigns a numerical value to rank similarity based upon a theoretical calculation of what amino acids are more likely to mutate over generations, thus introducing the concept of protein evolutionary scoring matrices. The protein scoring matrix originally used was the Percent Acceptable Mutation (PAM) matrix, which is a theoretical estimation of the rate of mutation at any given amino acid position, assigning a numerical value dependent upon the substituted amino acid's predicted effect on protein structure and function. FASTA is available at the FASTA server in the United States (http://fasta.bioch.virginia.edu/fasta_www2/fasta_list2.shtml) and at EBI (http://www.ebi.ac.uk/Tools/sss/fasta/).

10.3.4 The BLAST Algorithm

The fastest of the algorithms above, BLAST is also the least sensitive. BLAST is a *local alignment tool* that uses a two-step process to find very short regions of similarities between sequences. FASTA and BLAST are alike in that they find local areas of similarity. However, FASTA identifies clusters of similarities within a local region, while BLAST identifies an initial single very short area of similarity then seeks a second very short similarity area within 40 amino acids or nucleotides, upstream or downstream of the first hit. Compared to FASTA, BLAST locates considerably smaller areas of local similarity with lower *percent identities*. This is also the reason BLAST runs considerably faster than FASTA.

BLAST, as for FASTA, ranks similarity with numerical values. Protein BLAST makes use of a more extensive and robust scoring matrix, the Blocks of Amino Acid Substitution Matrix (BLOSUM). Unlike the PAM scoring matrix, BLOSUM is an empirically derived substitution matrix taken from the published literature of conserved blocks of protein sequences which were originally aligned and calculated from the now-retired BLOCKS database.[7]

Due to the fact that BLAST initially locates very short areas of similarity (which may or may not predict similarity along the entire length of the Query and/or Subject sequence), and in combination with the underlying mathematical and biological assumptions involved with the BLOSUM matrix, BLAST uses *probability statistics* to determine which hits it retrieves might be due simply to mathematical chance that they appear similar when they are not. It reports this value in the BLAST results as the *E-value*.

Versions of BLAST are available at many bioinformatics databases worldwide, particularly the major ones, as the sequence similarity search tool of choice. EBI also provides an extensive set of BLAST and BLAST-like algorithms (http://www.ebi.ac.uk/Tools/sss/), often with a wider range of adjustable options than found elsewhere.

10.4   HOW BLAST WORKS

Understanding how BLAST functions is paramount to making informed and sophisticated decisions when configuring searches and analyzing results. In preparation for the protein BLAST tutorial provided in this chapter, this section details the step-by-step process involved in the BLAST algorithm locating, calculating, retrieving, and reporting sequence similarity hits, noting which steps are customizable at the BLAST interface.

Step 1. Computationally preparing the Query sequence. BLAST first identifies the left-hand side of a query. For a nucleotide sequence, it

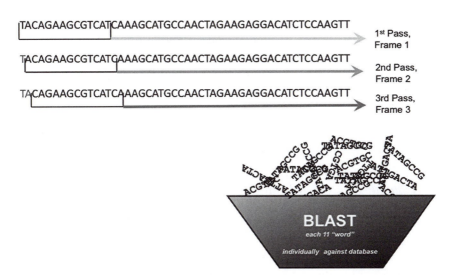

Figure 10.2 How nucleotide BLAST computationally prepares the query sequence to take in account nucleotide reading frames. When a nucleotide query is entered into the BLAST search box, neither the researcher nor BLAST, is able to identify which frame of three nucleotides is the correct reading frame for the encoded protein. To compensate for this, BLAST makes three separate passes down the query in all three reading frames. Beginning at the left-hand side of the query sequence, BLAST mathematically cuts the sequence into segments of 11 nucleotides called "words". It then makes a second pass, beginning one nucleotide in to the right of the first pass, which is the possible second reading frame. Finally, it makes a third pass two nucleotides in from the left-hand side of the query for the third reading frame. It then combines all segments from all three reading frames and continues its computations.

computationally "chops" up the sequence into units of 11 nucleotides beginning at the left-hand end (Figure 10.2) and sets them aside. It then moves one nucleotide in from the left-hand side of the query, "rechops" the sequence into 11 segments, adding these new "chops" to the first pass. It then makes moves two nucleotides in and performs a third pass of 11 "chops". If a protein sequence is the query, it performs the same three passes, but three amino acids at a time. The groups of 11 or three are called *Words* and are the default *Word Size* for nucleotide BLAST and protein BLAST, respectively. The Word Size can be adjusted to other values at the BLAST search interface.

Typically, sequences arrive at the BLAST query box direct from the research bench. For nucleotide sequences, BLAST cannot determine, nor does the researcher often know, which frame of three codons is read *in vivo* to generate the correct protein sequence. Therefore, BLAST reads all three possible coding frames for nucleotide queries. The default

Word size of eleven nucleotides at nucleotide BLAST represents three almost complete codons or three amino acids in a protein sequence. By setting the nucleotide BLAST default Word Size to 11 and the protein BLAST default Word Size to three, BLAST statistically handles nucleotide and protein sequences equivalently.

Step 2. IDENTIFYING LOCAL AREAS OF SIMILARITY. BLAST then takes all the Word Size "chops" it has generated from a Query sequence and runs them against the database selected by the researcher *(Subject* database), looking for areas of similarity on the *Subject sequences* held within the *Subject database*. When found, these areas of similarity are call *seeds*. BLAST will find many seeds along the length of the query sequence, either from within the same subject sequence or from different ones.

BLAST takes each seed found in turn and attempts to locate a second seed on the same *Subject sequence* that is within 40 base pairs (nucleotide query) or amino acids (protein query) of the first seed, either upstream or downstream. The distance between two seeds is called the *Window Size*. The *Window Size* of 40 is not customizable by the user, but is reported in the Search Summary section of every BLAST search.

If it finds a second seed, BLAST assigns a numerical score to the degree of similarity on the *Subject sequence* between the two seeds. Any alignment score with a numerical value of 11 (called *Threshold*) or above is considered a *Hit*. Anything less is discarded. BLAST assigns numerical scores differently for nucleotide and protein sequences, as discussed in Step 4 below. The *Threshold* is different for the various NCBI programs, and cannot be customized by the user. However, it is reported in the Search Summary section of BLAST search results.

Step 3. SEQUENCE ALIGNMENT. BLAST retrieves hits with numerical scores above the Threshold value and establishes an optimal residue-to-residue alignment between the query sequence and each subject sequence retrieved. As it finds the best alignment between the two sequences, it preserves both query and subject sequence order. To get the best possible alignment, BLAST slides the query and subject sequences against each other, artificially breaking apart (gapping) either the query or subject sequence to do so. This is illustrated in the example below, where the artificial gaps created during an alignment are shaded in grey:

Alignment possibility A:
ATGGCGT
* * * | | * * = +2+2+2+0-3+2+2 = +7
TAC—ACA
Alignment possibility B:
ATGGCGT
* | | * | * * = +2+0-3+2-3+2+2 = +2
T—ACACA

Figure 10.3   Assigning raw numerical values to rank nucleotide BLAST hits. After BLAST identifies local areas of similarity it aligns each database hit with the query sequence. To align each hit to the best possible similarity score, BLAST may have to create "gaps" in either the query or database sequence. It then assigns a numerical value of +2 to each nucleotide base-pair match, −3 for each base-pair mismatch and zero to each gap created along the entire query length. Shown are examples of BLAST finding two different hits in the database. Both examples show only a very short segment of the entire BLAST hit. In one case, the raw score of the partial hit totals to seven, in the other to two. This raw score is then used to calculate the *E*-value.

Step 4. ASSIGNING A RAW NUMERICAL SCORE TO EACH HIT. Once aligned optimally, BLAST ranks sequence similarity by assigning a *raw numerical score* to each sequence retrieved, as shown in Figure 10.3. For nucleotide sequences, BLAST assigns numerical scores as follows: +2 is given for each *exact* base pair match, −3 is given for each mismatch, and a default value of zero is given to any gaps, making them neutral in scoring. BLAST adds up the total numerical value along the length of the alignment and reports it as the *raw score*, as illustrated in Figure 10.3.

Assigning a raw score to protein sequence similarity alignments is more complicated than to nucleotide sequences as there are 20 amino acids compared to the four nucleic acid nucleotides. NCBI scoring matrices can be customized by researchers and include BLOSUM 45, 50, 62, 80 and 90 options. All of the BLOSUM matrices find common functions in sequences based upon locating common domains in a query sequence. In essence, BLOSUM scores what stays the same (is conserved) over evolution. The degree of sequence conservation depends upon the particular BLOSUM matrix chosen.

The default protein scoring matrix at protein BLAST is BLOSUM62 (Figure 10.4), where the number 62 represents numerical amino acid substitution values taken from a reference set of blocks of conserved sequences that amongst themselves are more than 62% *identical*. In a similar fashion, the BLOSUM 80 scoring matrix is derived from a conserved sequence set that is 80% identical, while BLOSUM45 is from

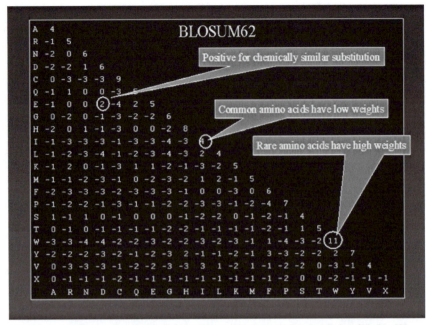

Figure 10.4 The BLOSUM62 protein scoring matrix. Assigning a raw score to protein BLAST involves assigning a numerical value for each and every amino acid substitution and its predicted impact on protein structure and function. To determine the score for any individual amino acid, find the amino acid on both the X- and Y-axis. As an example, tryptophan (IUPAC symbol W) carries a score of 11 when it is not substituted in the protein sequence. To find what impact the substitution of threonine (IUPAC symbol T) of tryptophan (IUPAC symbol W) has on calculating the raw score, first locate tryptophan (W) on the Y-axis, then read across and locate threonine (T) on the X-axis. The score drops to −2 when tryptophan is substituted with threonine. Rare amino acids in protein sequences such as tryptophan or proline, or amino acids which impart special function to proteins such as tyrosines, which are typically phosphorylation sites involved in important protein functions, carry large scoring penalties if substituted as they can impact protein structure and/or function significantly.

a reference set of sequences that are only 45% identical. Choosing BLOSUM80 as a scoring matrix would be expected to weight towards retrieving sequences that have higher similarity and can be predicted to be in closely related organisms as a result. Thus, if a human sequence was the query sequence, and only sequences from primates were desired, BLOSUM80 might be considered over the default BLOSUM62. In contrast, BLOSUM45 has the capability of pulling sequences that have diverged further from each other.

NCBI provides PAM protein scoring matrix options as well. In contrast to BLOSUM, PAM matrices score what changes over evolutionary time since they are based upon what is a permissible mutation rate for any given sequence to retain its function (see Section 10.3.3). Permissible mutation rates can exceed 100% as PAM adds up every single mutational event at any given residue (point mutations, insertions, deletions, *etc.*) over evolutionary history. NCBI offers three PAM matrix options: PAM 30, 70, and 250. PAM30 means that the scoring matrix permits 30 mutations per 100 amino acids of sequence, while PAM70 permits the higher mutation rate (and greater evolutionary distance) of 70 mutations per 100 amino acids. PAM matrices are more likely to return hits from a greater range of divergent sequences, or hits in organisms or genes at a greater evolutionary distance than BLOSUM can. If a research group was working on a bacterium in a clinical research environment and desired to locate the equivalent proteins in humans or mice, a PAM scoring matrix could be a better matrix to choose over the default BLOSUM62 matrix. Wheeler [8] has performed an in-depth analysis on what factors to consider when selecting NCBI protein scoring matrices beyond the BLOSUM62 default, including empirically determining the relationships between BLOSUM and PAM matrices. Figure 10.5 is a visual adaptation of those results.

Step 5. Raw scores are Then normalized and reported in BLAST results as Maximum Scores. Higher *Maximum Scores* reflect higher degrees of similarity. The *Maximum Score* is portable to BLAST searches run elsewhere, both at NCBI and worldwide. They can be used to identify the best BLAST resource for a particular sequence as well as to compile a ranked list of BLAST results for the same sequence from multiple databases.

Step 6. BLAST derives the *E*-value from the maximum score and reports it in the BLAST search results. Due to the fact that BLAST initially seeks very small local areas of similarity, as well as the assumptions inherit in assigning similarity through scoring matrices, the

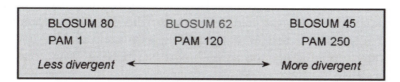

Figure 10.5 A guide to the interrelationship between BLOSUM and PAM matrices to aid in changing protein scoring matrix options during a BLAST search.

algorithm cannot guarantee that some sequence alignments are not due purely to mathematical chance. Therefore, BLAST mathematical calculations are based upon *probability statistics*. An *E-value* represents the "expectation" that a BLAST hit is NOT a random or chance similarity hit. This could be viewed as approximately equivalent to a false positive (experimental science/medical data) or as a "false drop" (library sciences). As an example, an *E-value* of 1 can be interpreted as meaning that one might expect to see one hit with a similar score simply by chance. *The closer to zero an E-value is, the less likely that a given hit is random and represents hits with true areas of similarity.* For this reason, BLAST loads its search results with those hits closest to zero first, with 0.0 being the "best" E-*value* score. There is an inverse correlation between *E*-values and Maximum Scores. The lower the *E-value*, the higher the Maximum Score will be. Unlike Maximum Scores, *E-values* are NOT portable to other searches or databases. Comparison of *E-values* can only occur from within the same set of search results.

The default *E-value* limit at NCBI BLAST is set at 10. At this setting, hits can be returned with *E-values* ranging from 0.0 (best) to 10 (worse). However, users can call out a specific beginning or ending *E-value* or a range of *E-values* from search results using the Reformatting Options tool (Section 10.9.1). Finally, it must be kept in mind when analyzing BLAST results that *E-values* report the probability of the degree of similarity between sequences, not necessarily the biological significance of those similarities, which is up to the user to determine in the context of the research project involved.

10.4.1 Implications for Determining Best Query Length

The longer the query length, the higher the Maximum Score will be and the best chance to statistically calculate the *E*-value. Investigators often have a choice of more than one query sequence to BLAST, either experimentally derived, multiple GenBank and/or 3D structure protein records of varying sequence length, or returned by organizational genome/protein sequencing services. Choosing the longest query length enhances recovering hits with higher degrees of similarity. When in doubt, the best query sequence length can be directly determined by comparing BLAST results between the available choices using the Edit and Resubmit function available at NCBI BLAST interfaces (see Section 10.10).

Minimum query length can be calculated from adding the Word Size used for the search plus two seeds of 40 each. This insures that BLAST has a chance of finding a match 40 amino acids (protein BLAST) or 40 nucleotides (nucleotide BLAST) on both the upstream and downstream side of the first seed found by BLAST. For protein searches using the default Word Size of 3, this would be at least queries of 83 amino acids (40 amino acids upstream + 3 Word Size + 40 amino acids downstream). For nucleotide BLAST, this value would be a query of 91 nucleotides using the default Word Size of 11. To hedge this slightly, a rule of thumb is to try to run BLAST searches with query sequence lengths of 120 residues or longer. BLAST will run at shorter lengths, but these values maximize the chance of BLAST finding multiple seeds in the query length.

Query lengths shorter than 40 nucleotides or amino acids, however, cannot be effectively run with the standard BLAST algorithms as they are too short for BLAST to locate the first seed. Fortunately, NCBI offers a modified BLAST algorithm to automatically adjust for short input search sequences of 30 nucleotides or amino acids (Section 10.11.1) insuring that users can work with very short sequences such a primers, motifs or epitopes of various kinds.

10.5 THE ART OF INTERPRETING *E*-VALUES

The BLAST algorithm was developed before the Human Genome Project, when bioinformatics databases were much smaller than today, records sparsely curated and, at NCBI, minimal inter- and intra-database hyperlinking from BLAST results to aid in analyzing search results for biological significance and functions. At that time, *E*-values represented a dominant mechanism through which to evaluate BLAST results. Today, databases are large, records can be extensively curated and NCBI's resources are extensively hyperlinked directly from individual BLAST hits. It is not unusual for BLAST results to contain many hits with the same or almost the same *E*-value. Due to the large size of databases today, most of these hits will probably be redundant: The same gene in closely related organisms and/or multiple hits to the same sequence held in many different GenBank records. Limiting BLAST results to a subset of organisms of interest will considerably remove unnecessary hits and make interpretation easier.

Nevertheless, it is important to keep in mind that an *E*-value of zero still has the chance that it is completely unrelated to the biological context that brought a researcher to run it as a BLAST query in the first place. Conversely, *E*-values approaching 1.0, or even greater, could indeed be valuable hits, particularly if the query and hit sequences are in organisms and/or genes that have diverged significantly over evolutionary time from the originating query's organismal source. *When evaluating which E-values have significance, it is best to keep in mind what organisms are involved and how far they have diverged in evolutionary time.* In addition, biological processes, as well as individual proteins can be under different evolutionary pressures. *What is found to be a predictive E-value for one protein may not be the same for other proteins within the same organism.*

Researchers with whom the author has worked with in the past decade are conservative in interpreting *E*-values. In general, an *E*-value of zero is regarded as something that should be investigated as it is unlikely to be random. *When comparing E-values among closely related organisms,* 1e-32 is generally the cut-off *E*-value considered to be non-random. Values over 1.0 may be random. Values over 10 are probably random. It is when attempting to rank *E*-values between diverging organisms and/ or genes, that *E*-value interpretation enters more the realm of an art or a philosophy than a science. The author was involved in consulting on a unicellular organism whose genome had yet to be sequenced. BLAST results were being returned with the lowest (best) *E*-values greater than 1.0. The research group later empirically determined that *E*-values of 3–4 from human sequences were most accurate in predicting their next round of wet-bench experiments. For another set of proteins from the same organism involved in a related project, *E*-values of greater than 1.0 in any primate sequence were found to be experimentally predictive.

It is for this reason that calling out a large number of BLAST results and taking the strategic approach to limit the results based upon organismal groups and/or a range of *E*-values can be a very efficient way to identify what sequences and *E*-value range make sense in the biological context and/or successfully predict wet-bench experimentation and is covered in Sections 10.9.2 and 10.9.3.

10.6 CREATING A SEQUENCE SIMILARITY ALGORITHM STRATEGY/PROTOCOL

BLAST searching should be viewed as an experiment, complete with controls, protocols and a means to verify and reproduce results. While BLAST remains the sequence similarity algorithm of choice due to its speed, sequence similarity tools other than BLAST have a definite place in the bioresearcher's bioinformatics toolkit. After running BLAST searches, consider extending search results with other algorithms. Suggestions include, but are not limited to:

- Use Needleman-Wunsch to visualize and confirm the degree of similarity along the entire length of a query sequence with sequences harvested from BLAST searches. It also independently verifies BLAST results before continuing to the next research step.

- Use Smith-Waterman as a screening method to identify sequences containing small regions of similarity, such as protein domains, either in the query sequence before running a BLAST or with sequences selected from BLAST results.
- Because FASTA is more sensitive than BLAST in identifying local areas of similarity with higher percent identities, it provides an independent mechanism to confirm BLAST results before taking the project to the next experimental stage.
- When BLAST results are unsatisfactory or returning a low number of hits, consider running FASTA and if necessary, Smith-Waterman.
- BLAST makes no claims to identifying all similar sequences. After running a BLAST search and identifying relevant hits, consider always running a FASTA search to identify possible sequences missed by BLAST.
- Consider beginning BLAST search sessions with a "control" sequence whose results are predictable and well known to the user. There are times when BLAST searches can be problematic.
- Using the techniques and/or strategies introduced in the exercises below, test and/or optimize BLAST searches by adjusting one parameter at a time. Then analyze BLAST search outputs as one would for any other wet-bench experiments performed to optimize assay conditions and to develop a standard protocol.

10.7   NCBI PROTEIN BLAST (BLASTP) HANDS-ON EXERCISE

The following hands-on protein BLAST exercise was originally based upon and modified from a real-life set of experiments. The research group was working with a ciliated protozoan, which is often used as a model system to study nuclear development due to the fact that one of its two nuclei is routinely destroyed and subsequently regenerated as part of the organism's normal life cycle. The research group had developed several temperature-sensitive mutants that could not progress beyond a particular stage of nuclear regeneration at the restrictive temperature. The group had discovered a set of genes that permitted nuclear regeneration to occur at the restrictive temperature and subsequently cloned them. They had no idea what the cloned genes were and turned to BLAST sequence searching to find similar sequences to aid in assigning a function to their cloned genes.

While performing the tutorial, be aware that NCBI regularly under-goes interface changes, often unannounced. Although every attempt has been made to work with the most stable components of NCBI BLAST for this exercise, it is possible that the interface may not exactly match the figures in this chapter.

10.7.1 Preparing to Search

10.7.1.1 Download and Install the NCBI Molecular Viewer, Cn3D.

1. At the NCBI home page, http://www.ncbi.nlm.nih.gov, click on Proteins in the left-hand Resource (A–Z) column.
2. Once the Proteins page is displayed, scroll to Tools and click on the Cn3D link.
3. Select the link for the appropriate computer operating system and follow instructions for installing the Cn3D4.3 Molecular Viewer. Installation will only take a few seconds and will automatically load and configure.

10.7.1.2 Create a My NCBI Account. My NCBI is a tool developed to manage and save search strategies, as well as individual records, from searches performed at all NCBI databases (PubMed, Gene, Structure, Protein, *etc.*), as well as at BLAST. It is available in the upper right-hand corner of almost all NCBI web pages, including the BLAST home page, which is where this exercise accesses it. If you click on the My NCBI hyperlink once logged in, you will note that access is gained to database searches and records, but not to BLAST searches. Access to saved BLAST searches is gained only through BLAST web pages.

If you have a My NCBI account already, login. If you do not, perform the following steps:

1. Migrate to the NCBI BLAST home page from your preferred search engine, or point your browser directly to http://blast.ncbi.nlm.nih.gov/Blast.cgi.
2. Click on the MY NCBI hyperlink in the extreme right-hand corner of the BLAST home page.
3. Click on "Register for an account".
4. Follow the instructions.
5. Make sure you are logged in to MY NCBI. The username just created will now appear in the MY NCBI box in the upper right-hand. If not, reload the web page.

Migrate to the BLAST home page. In the upper left-hand corner of the BLAST home page (http://blast.ncbi.nlm.nih.gov/Blast.cgi) are two tabs labeled: Recent Results and Saved Strategies. The Recent Results tab temporarily stores BLAST searches strategies as well as provides a hyperlink to saved search strategies. Actual BLAST search results are not saved. The search *strategy* is saved. Bioinformatics records are very dynamic, with records removed and added on a daily basis at NCBI. The BLAST algorithm itself is constantly tweaked. When a saved BLAST search is rerun, BLAST will deliver different results each time.

There will be times when EXACT BLAST search results need to be saved. Examples include documenting BLAST searches for publications or for patent applications. In order to save the EXACT results from a BLAST search, investigators must print the entire set of results and/or capture the results in their entirety with screen capture software that permits web page scrolling or programs that permit capturing web pages into PDF or similar documents.

If the need arises during the exercise, you can rerun the exercise from the Recent Results Tab. Alternatively, you can permanently save your search from the Recent Results tab to rerun another day. Managing and Saving BLAST results are covered in more detail in Section 10.8.1.

10.7.2   Configuring and Running Protein BLAST Searches

1. At the BLAST home page (http://blast.ncbi.nlm.nih.gov/), select "protein BLAST" from within the Basic BLAST section.
2. At the Protein BLAST home page, click "reset page", found in the upper right-hand corner.

   EXPLANATION: This returns the protein BLAST page to its default settings from previous BLAST searches that can be held in cache on the computer in use or even from servers elsewhere. *It should become habit to automatically reset a BLAST page before each and every new sequence BLAST search.*
3. Click on the Clear hyperlink between the Query box and the Query Subrange pull-down menus.

   EXPLANATION: This clears the Query Box of hidden commands that sometimes carries over from search to search. *This also should become an automatic habit before each and every new sequence BLAST search.*
4. Type the following GenBank accession record number into the BLASTp query box (Figure 10.6A): AAB46352.1, which is a human sequence.

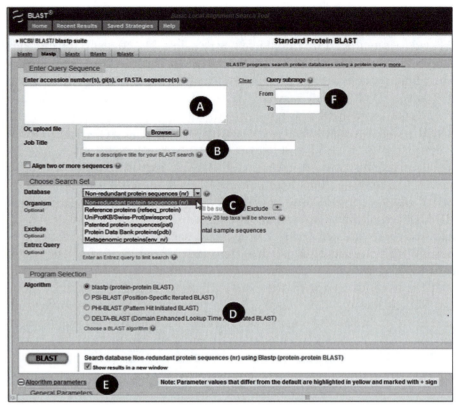

Figure 10.6   The protein BLAST (BLASTp) home page interface: (A) sequence query
box; (B) job title box to name searches; (C) database search option menu;
(D) protein BLAST algorithm options; (E) click open algorithmic parameters hyperlink to for options to adjust the BLAST algorithm itself; (F)
subrange query input where users can restrict BLAST to search a range of
residues within the sequence query. Clicking on the blue question mark
buttons next to each option will bring up a help file for that option.

EXPLANATION AND TIPS: Typing in an accession record number
acts as a proxy for the sequence held in the record. BLAST will
extract the record's sequence and use it as the query. In addition
to accepting NCBI accession record numbers, sequences can be
copy/pasted. BLAST recognizes most sequence formats including
FASTA, a raw or bare sequence, or sequences copy/pasted from
other NCBI resources or bioinformatics websites. You can also
use the browse button to upload a sequence file from your
computer.

5. Caution should be used copy/pasting sequences from word
   processor documents as they can hold hidden characters that

perturb BLAST searches. Prior to running BLAST or uploading a sequence file either convert sequences held in word processor documents to a .txt file format or use Notepad (Windows) or TextEdit (Mac).

6. Provide the Job Title "ChemPrimerBLAST1" (Figure 10.6B).

EXPLANATION: If no job title is provided, BLAST will assign a default generic, uninformative title from which you will be unable to distinguish your various BLAST searches from each other. Care should be taken in developing strategies to routinely assigning job titles that will distinguish the various searches from each other. Providing an organism name, laboratory unique experimental identifier, and the parameters of any search restrictions are highly recommended. An example job title would be "KU80_HUMAN_MaskedRepeatsSNP + GenBankLimit".

7. Choose a database (dataset). For this exercise, select the default non-redundant protein sequence (NR) dataset (Figure 10.6C).

EXPLANATION AND TIPS: The various protein and nucleotide datasets, along with brief descriptions, are provided at the BLAST Help Tab at the top of every BLAST web page. After clicking on the Help Tab, select the BLAST Program Selection Guide hyperlink. At the next web page, select the BLAST Database Content.

NR is the largest dataset available in protein BLAST. Providing a "big picture" view of related proteins and their functions, it should be the first dataset always run with a protein sequence. Although many BLAST users never BLAST against any dataset other than the default NR dataset, it is worthwhile considering BLASTing against the more specialized options available after first becoming familiar with the results returned from the NR dataset. As examples:

a. If you are interested in returning BLAST hits for 3D protein structures, consider selecting the Protein Data Bank dataset, which only holds 3D sequences. Three-dimensional protein structures at NCBI are sparse. Choosing this dataset is an efficient and effective way to rapidly recover 3D structures with sequence similarity to your query sequence.

b. BLASTing against the patent dataset will retrieve related sequences from United States patents granted to sequences held in GenBank records. Links are provided to the full-text of the patent. This dataset could be considerably instructive in helping to assign function to sequences or recovering similar sequences to engineered proteins. Patents can contain pro-prietary information related to protein structure, function and mutational analyses not found in the published literature.

8. Choose a protein BLAST algorithm (Figure 10.6D). For this exercise, make sure the default blastp algorithm is chosen.

    EXPLANATION: The protein BLAST algorithms available at NCBI are blastp, PSI-BLAST, PHI-BLAST and DELTA-BLAST. BLASTp, the default algorithm, is the standard BLAST program for running all protein BLAST searches. PSI-, PHI- and DELTA-BLAST are specialized algorithms for specific purposes. NCBI provides a guide to help users chose between the various BLAST programs. To access it, click on the Help Tab at the top of the page and then click on the "BLAST Program Selection Guide" hyperlink. At the next web page, select the "Program Selection Table" hyperlink.

9. Make sure the "show results in new window" box is checked (Figure 10.6E).

    EXPLANATION: This preserves the original BLAST query page, in its original browser tab, from being overdrawn by the BLAST search results, enabling you to efficiently return to the BLAST search without losing the original search para-meters you set.

10. Click the "algorithm parameters" hyperlink at the very bottom of the protein BLAST web page (Figure 10.6E). For this exercise, pull down the menu next to "max target sequences" and select: 1,000.

    EXPLANATION AND TIPS: The default value, 100 hits returned, is too small to yield significant results. For actual experimental BLAST searches, the number of return hits to specify will depend upon the quantity and quality of the BLAST results. Selecting the maximum number of 20,000 hits available for an experimentally-derived sequence run for the first time in a BLAST would not be unusual. This may seem like a large number of records to evaluate, but by using the BLAST Formatting Options in this exercise, you will be able to filter and limit the results repeatedly, permitting evaluation and analyses on much smaller subsets of records without having to rerun BLAST.

    The Algorithm Parameters of the BLAST search interface also permits tweaking of the BLAST algorithm itself (Figure 10.7):

    a. Changing the default E-value maximum cut-off limit of 10, the default BLOSUM62 scoring matrix, and the default word size of 3.

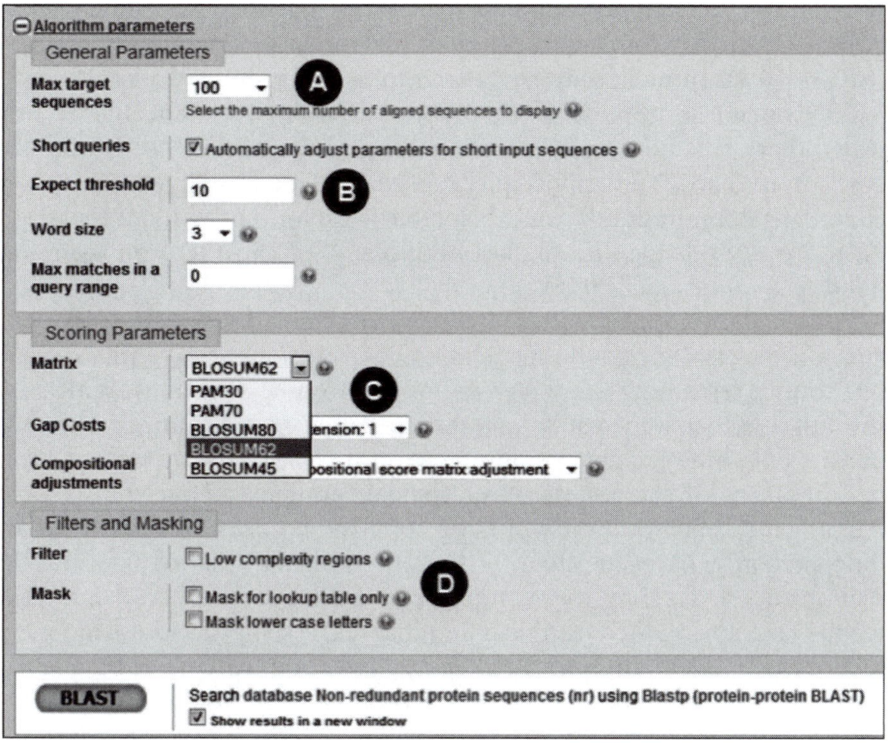

Figure 10.7 Adjusting the default BLAST algorithmic parameters: (A) pull-down menu to choose total hits returned from a BLAST search; (B) *E*-value cut-off threshold, when set at 10 BLAST will return hits with *E*-values up to and including 10; (C) scoring matrix options; (D) filter or "mask" residues to instruct BLAST to ignore them. Clicking on the blue question mark buttons next to each option will bring up a help file for that option.

    b. Altering gap size and extension should be left to advanced users with considerable experience as small changes can have profound effects on BLAST results.

    c. Ability to manipulate portions of the query sequence that BLAST should ignore, by "masking" sequences within the query (Section 10.11.3).

11. For this exercise, leave all other options at their default settings.

12. Click on the blue BLAST button to run the BLAST search.

10.8   UNDERSTANDING BLAST RESULT SECTIONS

It may take several minutes for the BLAST search results to fully display. When BLAST locates sequences matching a query from within the

NR dataset, it will immediately search for and display a hyperactive graphical display with any known protein domains in your query (Figure 10.8). In most searches, the graphical display will automatically load. Depending upon individual computing environment and/or fire-walls, there will be times that a button must be clicked to display the graphic instead. The input query itself will be displayed above the conserved domain graphic as a black bar with amino acid 1 of the query on the left-hand side and the last amino acid of the query on the right-hand side (amino acid 996).

After the Conserved Domain graphic loads, BLAST will continue working in the background to retrieve and organize all hits found, refreshing the screen from time to time as it does. Wait until the full 1,000 search results are retrieved before continuing. This will consist of a blue header containing several hyperlinks, as well as three permutations of the results: A Graphical Summary; Descriptions; and Alignments. Each section will have a button to click to open or close independently of each other. The default display is to have all the sections open. To simplify the interface for this exercise, scroll down the results list, closing the Graphical Summary, Descriptions and Alignment sections as shown in Figure 10.9. Each section will be subsequently reopened and explained in detail separately later in the exercise.

10.8.1 Saving BLAST Search Strategies

1. Open the Recent Results Tab at the top of the web page.
2. Locate the ChemPrimerBLAST search.
   a. Note that BLAST provides a finite expiration date for the search.
   b. Clicking on the Request ID hyperlink will immediately reload the BLAST results in the browser's window.
3. To permanently save the search, click on the Save hyperlink.
4. Migrate to the *Saved Strategies* Tab. The search is now perma-nently saved.
5. To rerun a saved search to return to this Primer if necessary, click on the *Saved Strategies* Tab.
   a. Click on the "View" hyperlink corresponding to the search you want to rerun.
   b. BLAST will reload the search parameters as already defined.
   c. Click on the BLAST button to rerun the search.

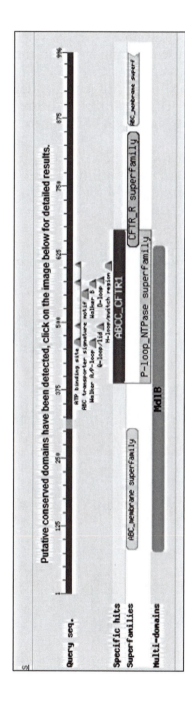

Figure 10.8 Assigning protein function to a protein sequence through the use of the interactive conserved domain display. BLAST will automatically identify any conserved protein domains within a query sequence as part of its algorithmic process. In this figure, the 996 amino acid query is displayed at the top with query residue number 1 at the left-hand side. If putative conserved domains are found, BLAST aligns the domains to their point within the query sequence. The query displayed contains two ABC membrane protein superfamily common sequences, a Md1B domain, the ABCC_CFTR1 domain, and a CFTR_R protein superfamily common domain. In this search, BLAST was able to locate and map specific functional sequences of the ABCC_CFTR1 domain to specific points on the query sequence. These include 4 distinct possible ATP binding sites between query residues ~460–610, a single ABC transporter signature motif at approximate query residue 550 and several other protein loops with known function elsewhere within the sequence. The conserved domain display is interactive. Clicking on the various domain images will take the researcher to detailed information held in the NCBI Conserved Domains database.

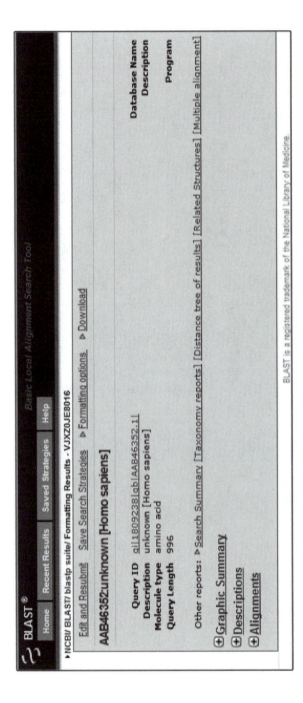

Figure 10.9 BLAST results are quite extensive, separated into collapsible sections titled Graphic Summary, Descriptions and Alignments. Collapsing the sections, as shown here, permits easier analysis and migration through the results. Note that at the top of the results are hyperlinks for editing and resubmitting a BLAST search, saving searches, filtering and sorting results through the formatting hyperlink, and a link to download results.

10.8.2 BLAST Results Header Section

Across the top of the blue header section (Figure 10.9) are the Edit and Resubmit (running new BLAST searches), Save Search Strategies, Formatting Options (temporarily limiting search results, or changing how the results are viewed) and Download hyperlinks. The center section generally summarizes information about the input query and the dataset against which it was run. At the bottom of the header are the "Other Reports" hyperlinks: Search Summary, Taxonomy reports, Distance tree of results, Related Structures and Multiple Alignment.

1. Click the Search Summary hyperlink.

   The Search Summary details the parameters of the search, the dataset queried, and the specific BLAST statistics used to run this search. Note that it contains settings for the Word Size, Threshold, scoring matrix, and the Window Size. (See "How BLAST Works", Section 10.4, for a review of terms.) Although the Word Size can be changed by the searcher, Threshold and Window Size cannot. For researchers who need to preserve this EXACT set of results, saving this information at the time of the search is necessary, as future BLAST versions may calculate these parameters differently.

2. Click the Taxonomy Reports hyperlink.

   Taxonomy Reports is a hyperlinked hit list of results by organism. It is very useful to screen this information for the range of organisms containing a similar sequence to a query, as well as evaluating whether this search is yielding sufficient hits in the organism(s) of interest.

3. Click the Distance tree of results hyperlink.

   The Distance tree of results complements the Taxonomy reports. While the latter lists organisms by taxonomy (phyla), the Distance tree organizes the sequences found by their "genomic" or evolutionary distance (phylogenomics).

10.8.3 BLAST Results Graphic Summary Section

Click on the Graphical Summary link (Figure 10.10). By default, BLAST loads approximately the first 100 hits in the Graphical Summary section. To fully display all 1,000 hits requested:

1. Scroll to the top of the BLAST page.
2. Click on the Formatting Options link. The full Formatting Options window will open (Figure 10.11).

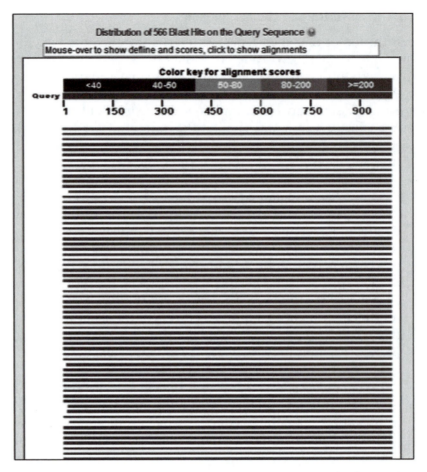

Figure 10.10 The Graphical Summary portion of a BLAST result. BLAST displays results in a variety of modes including a graphical display. The query sequence from 1–>900 residues is at the top. Red denotes that at least 200 residues are similar in the hit, green 50–80, blue 40–50, *etc*. In this search, most of the hits returned are greater than 200 residues in common with the database hit and extend the entire length of the query. A few hits begin several residues internal to the left-hand side of the query sequence. The full results are not shown here. This search also yielded hits in blue 40–50 range predominantly between the 450th and 600th query residue and red hits restricted to shorter regions throughout the query sequence.

3. In the *Limit* results section of the Formatting Options window, pull down the menu next to the "graphical overview" and select 1,000 to match the total number of BLAST hits called for in the search.

4. Click on the blue Reformat button in the upper right hand corner. The BLAST Graphical Summary will redraw with the full complement of the 1,000 hits.

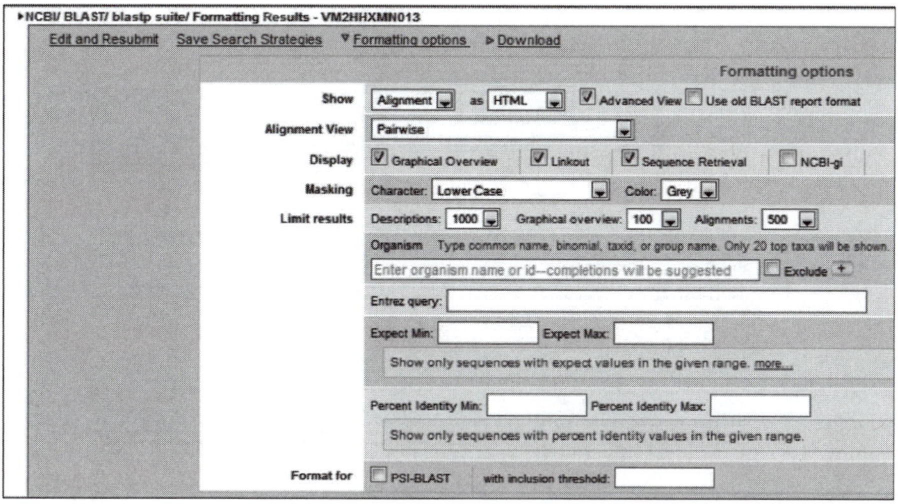

Figure 10.11 The BLAST Formatting Option box. BLAST permits researchers to temporarily filter or sort results to make analysis easier. Among choices, users can change how alignments are viewed, restrict searches to specific organisms, and limit the *E*-value and percent identity through the use of the Expect. Min and Expect Max or the Percent Identity Min and Max input boxes, respectively.

5. Mousing over the Graphical Summary window displays the accession number for each hit in the text box at the very top of the Graphical Summary.

6. Each hit is color coded, with red representing a better alignment to the query sequence, then purple, then green, then blue, then black.

7. The length of the hit indicates the per cent coverage in the query. A hit extending across the width of the Graphical Summary indicates it spans the entire length of the input query (100% query coverage).
   a. Where there is no hit to a region of the query sequence, the Graphical Summary will be blank.

8. Clicking on one of the hits in the Graphical Summary takes you to the corresponding alignment in the Alignment view further down the page.

9. The order of the hits in the Graphical Summary does not necessarily represent the order of hits in the Descriptions and Alignment result sections. BLAST squeezes the hits into the most efficient visual display. A hit from one gene may lie next to a hit from an unrelated gene. Figure 10.12 provides a visual guide to interpreting how NCBI organizes the hits within the Graphical Summary.

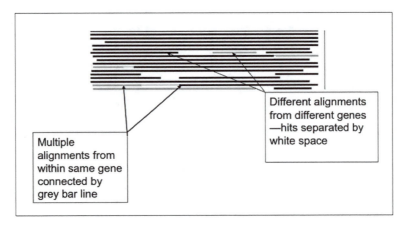

Figure 10.12   Interpreting the BLAST Graphical Summary hits and their relationships to each other.

10.8.4   BLAST Results Descriptions Section

Close the Graphical Summary section and Open the Descriptions Section (Figure 10.13).

1. Presented as a sortable table, the Description contains the 1000 user-specified hits through the Maximum Target Sequences. The default sort is by *E*-value, with the best *E*-value (closest to zero) presented first. Clicking on the *E*-value column header reverses the sort; a function that works on all column headers.
2. The right-most column contains the accession record number for each hit found. Clicking on it opens the actual record.
3. The Alignment column provides the title from the accession record. A rapid scroll through the titles is informative. This sequence may be similar to the cystic fibrosis transmembrane regulator protein and/or the ABC transporter protein family, may have a function in multi-drug resistance in some organisms, and seems to be found in many different organisms, particularly eukaryotes.
4. The Maximum Score column is the normalized alignment score BLAST used to assign numerical values to each hit while calculating the *E*-value. (Refer to Section 10.4 for review). It can be used to compare BLAST searches between bioinformatics resources worldwide.
5. The Query Coverage column specifies what percentage of the query is aligned with a hit. A value of 100% indicates the hit spans the length of the input query.

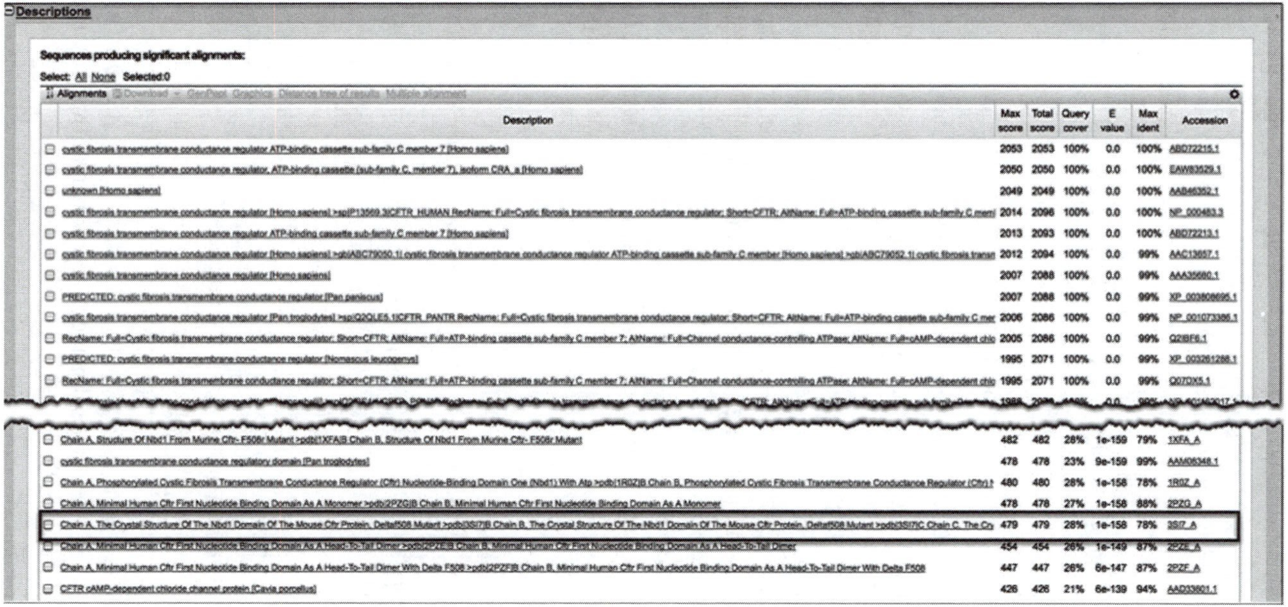

Figure 10.13 BLAST results as displayed in the Descriptions Table. The right-hand column contains the accession numbers for each hit returned. If the hit is in a NCBI sequence record, the NCBI accession record number is returned (*e.g.*, AAG13656.1, NP_001073386.1). If the hit is in a 3D structure record, the Protein Data Bank accession record number is returned (*e.g.*, 1XFA_A, 1R0Z_A). A unique feature of NCBI Protein BLAST is the ability to view any BLAST hit to a retrieved protein 3D structure. Note that the hit for the PDB 3D structure record 3S17_A (see section 10.8.5.1) in the boxed-in frame. Clicking on the URL for the title to this record (Chain A, The Crystal Structure of the NBD1 Domain of the Mouse CFTR protein. . . .), will open the 3S17_A PDB Structure record and its pre-computed alignment to the BLAST query sequence hit.

6. The *E*-value column ranks each hit's probability that it is a random hit, with zero being the least chance of randomness. Another way to state this is that the lower the *E*-value, the more significant the alignment. Unlike the Maximum Score, *E*-values can only be used for comparative value *within* the same set of search results.
7. The Maximum Identity column specifies the percent similarity (identity) between the query and a hit. Thus a hit could span 72% of the query (Query Coverage column), but be only 42% similar.

10.8.5   BLAST Results Alignments Section

Close the Descriptions Section and open the Alignments Section of the results. The Alignment section displays the actual alignments found by BLAST for each hit retrieved from the NR dataset to the input query, where Query is the input sequence and Subject is the sequence retrieved from the BLAST search. The alignment is given in batches of 60 query amino acids at a time, beginning at amino acid 1 of the query sequence. For each amino acid in the alignment, the identical, conservative, mismatch, and gap substitutions are indicated. As shown in Figure 10.14, identical amino acid matches (*i.e.*, leucine for leucine) are identified by the amino acid involved, while conservative substitutions (substitutions such as valine for alanine) are denoted with a plus (+) sign to indicate

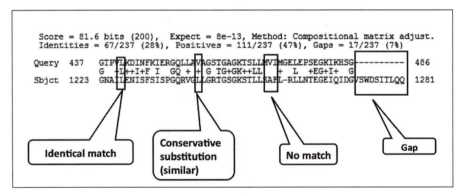

Figure 10.14   Interpreting BLAST results with the Alignment view. BLAST identifies "similarity" as both identical matches and conservative substitutions. Conservative substitutions are denoted by a + symbol. Several styles of alignment views are available to choose from through the Formatting Options box. Shown here is the default "pairwise" alignment view. Refer to section 10.9.1 for more details on the various alignment view options.

similarity in the sequence. Any gaps created by BLAST during optimum alignment are shown with a dashed line.

10.8.5.1   Analyze Biological Function by Navigating Laterally to Other NCBI Resources.   Linking laterally through NCBI to these major resources is critical to assigning a function to the sequences retrieved by BLAST. Practice lateral linking by clicking on any Structure, PubChem, MapViewer, or Gene link under the Related Information section to the right of each alignment in the Alignments section.

1. STRUCTURE LINKS: Any BLAST hit with a Structure link indicates that the hit lies in a 3D protein sequence. Clicking on the Structure link will load the NCBI Structure record *upon which the query sequence has been pre-BLASTed for analysis* (Figure 10.15).
   a. Locate the 3D structure record for 3SI7_A, scrolling the alignments list while searching for its *E*-value of 1e-158 in the *E*-value column (Figure 10.13). Once at the link, click on the URL for the title of the record (Chain A, The Crystal Structure of the Nbd1 Domain of the Mouse...).
   b. Click on the Structure link on the extreme right-hand side of the alignment display, under Related Information. The 3SI7_A 3D Structure record will be retrieved.
   c. At the top of the Structure record page is the domain graphic. Underneath it is a pink bar (Figure 10.15A) indicating where the query aligns to the 3D structure.
   d. Below that, the alignment is shown residue by residue with the BLAST Query sequence on top and the Subject 3SI7_A 3D structure sequence beneath (Figure 10.15B). Red residues are aligned residue hits identified by BLAST as either an exact match or as a conservative amino acid substitution (*i.e.*, "similar"). Residue mismatches are in blue.
   e. To see the BLAST hit aligned on top of the 3S17_A 3D structure in three dimensions, click on the button "View Structure and Alignment in Cn3D". The file will download to your computer and automatically launch the Cn3D Molecular Viewer that was installed at the beginning of this tutorial. If Cn3D is not installed, you will be queried whether to install it or not. Alternatively, a manual install can be performed by clicking on the hyperlink below the "View Structure" icon. Cn3D displays the 3S17_A structure. Wherever 3S17_A is similar in sequence to the Query hit, it is colored red. Wherever it is mismatched to the query, the 3S17_A residues are colored in blue.

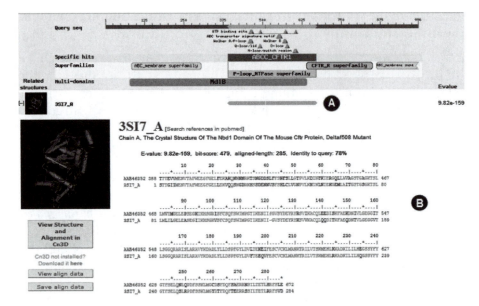

Figure 10.15 Interpreting BLAST results through lateral navigation to 3D structures. If available for any given hit, NCBI provides icons in the Description Table for searchers to migrate to related NCBI resources. If BLAST locates a hit that is also in the 3D Structure NCBI database, BLAST will automatically BLAST the query sequence, on-the-fly, on top of the 3D structure. Clicking on the Structure icon will display both the aligned sequences, with similar residues colored in red and mismatches in blue, either as a sequence or as 3D model. (A) Denotes that part of the query sequence that is aligned with the 3D structure retrieved by BLAST. (B) The structure sequence hit aligned with the query sequence. Cn3D displays the structure sequence (3S17_A in this example) with residues colored red to show similarity to the query sequence and blue to show mismatches.

2. PUBCHEM LINKS: Clicking through to PubChem will load records of small molecules and that interact with any given hit sequence from the PubChem BioAssay database.

3. MAPVIEWER LINKS: Note that clicking through to MapViewer will position the hit on the appropriate chromosome, provide what DNA strand it is on, direction of transcription, what genes are nearby and provide access to the 5' upstream regulatory regions, and much more.

4. GENE LINKS: Click on the Gene link for the result with accession number ABD72215.1, which is the human CFTR gene. This will take you to the NCBI Gene Record for the human sequence, Gene accession number 1080.

Gene is a major resource at NCBI. Not a database but an index of hyperlinks, similar in function to a back-of-the-book-index, Gene collects in one place the current state of knowledge about a gene, its products, and their functions. There will be one record for each gene/organism. Most importantly, Gene is integrated to both NCBI resources and to external, third-party resources through an extensive set of hyperlinks down the right-hand side of the web page (Figure 10.16), divided into several categories.

- The upper "Table of Contents" region contains internal links to the various sections within the Gene Record itself.
- The middle "Links" section consists of hyperlinks to CFTR records found elsewhere at NCBI (Figure 10.16).
- Scrolling down, the last section is the "Links to other resources" which are authoritative third-party CFTR resources and tools.

Critical to assigning biological function(s) to retrieved BLAST sequences, Gene should be considered as the first click out of a BLAST hit result. Learning the purposes and functions of the various databases and resources at the other end of Gene's right-hand link list is an essential step to efficiently analyzing BLAST results.

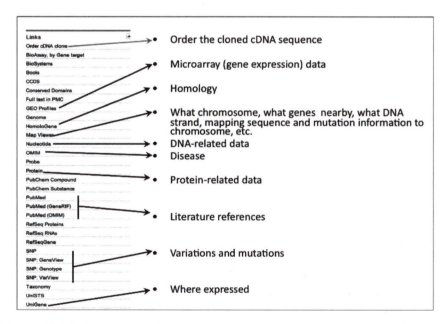

Figure 10.16 Interpreting BLAST results through lateral navigation to NCBI Gene. If available for any given hit, NCBI will provide a Gene icon in the Results Alignments section for searchers to migrate to the Gene Database. Shown here is the extensive "right-hand link" list of a Gene record for a BLAST hit.

10.9   MANIPULATING BLAST RESULTS

10.9.1   Reformatting the Alignment View

Several features of the search results can be temporarily reformatted or sorted using the "Formatting Options" box. One of them is the alignment view which currently is rendered in the default "pairwise" view as shown in Figure 10.14. To change the alignment view, perform the following steps:

1. Scroll to the top of the search results to the Formatting Options box.
2. Change the alignment view from "pairwise" to "pairwise with dots for identities".
3. Reformat by clicking on the blue Reformat button in the upper right-hand corner of the Reformatting Options box.
4. Scroll down to the alignment search results section. You will now see that the alignment view has been altered. Wherever there is a dot, the query and retrieved sequence are either identical or similar. Mismatches are identified in red.
5. There are actually six different alignment views available through the Formatting Options depending upon what analysis needs to be done and/or personal preference. Reformat for each one in turn, and then select your preferred alignment view for the rest of this exercise.

10.9.2   Reformatting (Limiting) BLAST Results to Organisms

The Formatting Options box (Figure 10.11) serves a dual function in transiently applying limits to an active BLAST search.

1. In the Organism text box of the Formatting Options, simply begin typing the name of the organism to which you would like to limit the search. NCBI will display a pop-up menu of up to 20 search possibilities from which to select, changing the options as you type. For this exercise, type "human" and select the human "taxid 9606" from the pop-up box of possible search terms that appear.
2. Click the Reformat button.
3. Only the human records retrieved by BLAST are now displayed.
4. Return to the Formatting Options box. Clear the Organism box by highlighting and using the computer keyboard's backspace key.
5. Re-format. The 2869 BLAST results return from the original search.

6. To limit to another set of organisms, return to the Formatting Options box.
   a. Enter Human into the limit box as before.
   b. Click on the + sign to the right of the box.
   c. A second organism search box appears.
   d. Type in *Drosophila*, choosing *Drosophila melanogaster*, taxid 7227.
   e. Click Reformat.
   f. The original BLAST search results have now been limited to records that are from EITHER Human OR *Drosophila*.
7. Return to the Formatting Options box. Clear both organism search terms. Reformat to redraw the complete original BLAST search. To exclude particular organisms from the search results, do the following:
   a. In the organism search box, type in vertebrate, selecting "Vertebrates, taxid 7742".
   b. Click the + button. Type in rat, selecting the Norway rat, taxid 10116.
   c. Check the Exclude box
   d. Reformat. Your BLAST results have now been limited (filtered) to all vertebrates except the Norway rat. This is the equivalent to the search string: Vertebrates NOT Norway rat.
8. Clear all the search boxes and reformat to bring back the original BLAST search.

   EXPLANATION AND TIPS: One of the most valuable BLAST strategies is to limit BLAST results to an organism or set of organisms. This creates a manageable subset of records for analysis or to sequentially analyze results on different sets of organisms. NCBI also permits limits to phyla groups. Either scientific or common names can be used. These include "mammalia", "flowering plants", "Neanderthal", "gram-positive bacteria" and more. The NCBI organismal concept also encompasses synthetically or artificially derived sequences (*e.g.*, "artificial gene", "synthetic construct"). When unsure of what search term might be possible, simply begin typing. NCBI will contextually display 20 choices in the pop-up box, changing as you type the term.

10.9.3   Reformatting (Limiting) *E*-Values

Perform the following steps to limit search results to a range of E-values between 1e-70 and 1e-76:

1. In the Formatting Options window locate the "Expect Min" and "Expect Max" text boxes.

2. In the left-hand Expect Min box, enter 1e-76 (the value closer to zero or the "better" *E*-value) and in the right-hand Expect Max box (the number closer to 1.0 and the "worse" *E*-value), enter 1e-70.
3. Reformat. The hits bounded between these two E-values are returned.
4. Clear the range by deleting the text and reformatting.
5. The original BLAST results are returned, once again.

EXPLANATION: With experience, researchers develop a concept of what range of *E*-values provides the best focus for analyzing BLAST results. Formatting Options can also be used to sort search results for a range of *E*-values. This permits analysis of BLAST results in manageable "chunks", as well as provides the strategy to identify appropriate *E*-value limits to set when running future BLAST searches for any given sequence. Although not covered in this tutorial, note that Formatting Options can also temporarily limit to ranges for percent identity. A range does NOT have to be specified. Either a minimum or maximum percent identity or *E*-value can be supplied.

10.10   EDITING AND RESUBMITTING BLAST SEARCHES

The Edit and Resubmit feature of BLAST is used when either a new sequence is to be searched or when BLAST parameters are adjusted on an existing search. It is an efficient way to experimentally compare (optimize) searches against different subject databases, determine the effect of, or experiment with, altering Threshold limits, which protein scoring matrix or set of matrices is best, switching between various protein or nucleotide BLAST algorithms, *etc.*

For this exercise, the ChemPrimerBLAST1 search will be reloaded and then edited using the "Edit and Resubmit" BLAST function to run the search against the ProteinDataBank database.

1. Reload the ChemPrimerBLAST1 search from the Saved Search Strategies Tab if it is not currently active.
2. Once the results are loaded, click on the Edit and Resubmit hyperlink in the upper left-hand corner of the web page.
3. Reset the page using the link in the upper right-hand corner.
4. Click the "clear" link to make sure the query box is free of any characters. Make sure that all parameters return to their default values and that there are no yellow highlighted areas in the interface. Repeat if necessary. If this still does not

return the interface to the default values, quit the browser and restart it.

5. At the Database pull-down menu (Figure 10.6C), switch from the NR dataset to the Protein DataBank (PDB) dataset.
6. Give the Job Title: ChemPrimerBLAST_PDBdatabase.
7. Leave all other default options the same from the original ChemPrimerBLAST1 search.
8. BLAST.
9. Click on the Edit and Resubmit hyperlink.
10. Switch from the PDB dataset to the patented proteins dataset.
11. Give the Job Title: ChemPrimerBLAST_Patent.
12. Leave all other default options as before.
13. BLAST.

   EXPLANATION: BLASTing against the patented sequences datasets runs a similarity sequence search against those sequences held in the patent division of GenBank, providing access to the full-text of the patent.

14. In the left-hand column of the Graphical Summary table, click on the accession record (ACM85813.1) hyperlink from the first hit.
15. Once at the record, locate the Journal field of the record. Click on the US Patent number hyperlink. You are now at the full-text of the patent at the USPTO full-patent database.

The Edit and Resubmit also has several features in common with the Formatting Options box, including limiting searches to organisms. Unlike the Formatting Options, however, Edit and Resubmit, *permanently limits* a BLAST search to the organismal limits set. To remove the organismal limits after the BLAST search is preformed, click on the Edit and Resubmit link, clear the organismal search set and rerun BLAST.

10.11   ADVANCED BLAST TECHNIQUES

10.11.1   Working with Short Queries (Primers, Motifs, Epitopes)

Both protein and nucleotide BLAST provide the opportunity to run short input search queries less than 30 residues in length.

1. Reload the ChemPrimerBLAST1 search from the Saved Search Strategies Tab if it is not currently active.
2. Scroll to the Algorithm Parameters. In the Short Queries section, make sure the box next to the option "automatically adjust parameters for short input sequences" is checked. Normally on by default, there are times when it may not be checked.

3. Reset the page using the link in the upper right-hand corner.
4. Click the "clear" link to make sure the query box is free of any characters. Make sure that all parameters return to their default values and that there are no yellow highlighted areas in the interface. Repeat if necessary. If this still does not return the interface to the default values, quit the browser and restart it.
5. Type into the search box in upper case letters: ELVIS.
6. Give the search the job title: Is Elvis at NCBI.
7. Select 500 maximum target sequences in the algorithm parameters section.
8. Leave all other parameters at their default values.
9. BLAST.
10. When results are returned, reformat the results to display 500 alignments using the Formatting Options box.

    EXPLANATION: Note that, when the BLAST results page draws, NCBI informs the searcher at the top of the page that "Your search parameters were adjusted to search for a short input sequence." Because this query is less than 40 amino acids in length, the graphical display section will always return the hits as a black bar even when *exact matches* are returned as in this case.

11. Scroll to the alignments section of the results and note that BLAST is returning *exact matches* (no substitutions) for ELVIS.

    EXPLANATION: This is the one algorithm at NCBI that is capable of finding exact matches to biological functional sequences including primer, motif, immune epitopes, drug epitopes, protein-protein binding sites, nucleic acid-protein binding sites, *etc*. Although in this example, all sequences retrieved are exact matches, this will not necessarily be true for research-derived sequences.

12. Click on the Edit and Resubmit link to run a new sequence query.
13. Reset page and clear query box as for steps 3 and 4, above.
14. Type into the search box in upper case letters: ELVISHA-SLEFTTHEROOM, which is 19 amino acids long.
15. Give the job title: Has Elvis left the room.
16. Select 500 maximum target sequences as before.
17. Leave all other parameters set to their default values.
18. BLAST.
19. When results are returned, reformat the results to display 500 Alignment and 500 Graphical Summary using the Formatting Options box.
20. Reformat.

EXPLANATION: When BLAST returns the results note that, in addition to the message that BLAST adjusted for a short input sequence, two new informational messages are returned:

- *Informational Message: One or more U or O characters replaced by X for alignment score calculations at positions 16, 17.*
- *Informational Message: Query 'lcl/95519 unnamed protein product' (# 1): Warning: One or more U or O characters replaced by X for alignment score calculations at positions 16, 17.*

NCBI BLAST makes use only of the single letter IUPAC codes for amino acids and nucleotides (http://www.bioinformatics.org/sms2/iupac.html). In this query sequence U and O are not recognized by BLAST and are replaced by X indicating that BLAST accepted substitutions of any amino acid at those positions. Scrolling to the alignment view of the results section clearly displays this. Compared to the shorter search above, a considerable amount of substitutions is evident in the returned sequences. The same phenomenon will occur when input queries contain symbols representing nonstandard amino acids or nucleotide, whether naturally occurring or synthetic. As a result, BLAST currently cannot recognize non-standard or modified amino acids or nucleotides within sequences.

10.11.2 Comparing (Aligning) Sequences

All versions of NCBI Basic BLAST can compare sequences, by aligning two or more sequences to each other. This feature can be accessed directly from any BLAST search page or from the BLAST home page in the "align two (or more) sequences" option in Specialized BLAST section at the bottom of the page.

1. Click the Edit and Resubmit link to run a new protein BLAST query.
2. Reset the page and clear the query box.
3. In the query box, type in the following sequence in upper case letters:
   a. CSQFSWIMPGTIKENIIGVSYDEYRYRSVIK
4. Give the Job Title: MutationCheckCFTR
5. Directly below the query box, check the box titled: align two or more sequences.
6. A second query box will open. In this box, input the accession number: NP_000483.3.
7. Leave all other parameters at default value.

8. BLAST.

EXPLANATION: A single hit is returned, aligning the first query sequence, the experimentally-derived sequence of interest to the researcher, to a chosen Subject sequence, in this case the full-length wild-type protein to the first sequence. To view the alignment better, reformat to display the alignment view as "pairwise with dots for identities". It is now clear that the Query sequence contains a deletion of phenylalanine compared to the wild-type, normal protein.

Multiple comparisons can be made simultaneously for either the Query Sequence, the Subject Sequence or for both. When inputting more than sequence into either or both query boxes, input a single sequence per line, beginning a new line with each additional sequence.

10.11.2.1 Aligning Primers to a Nucleotide Sequence. The compare (align) feature will work with aligning a primer to a nucleotide sequence at nucleotide BLAST. However, the default Expect Threshold should be increased from 10 to at least 1,000 with or without decreasing the default Word Size. Unchecking the boxes in the filter section for Low Complexity regions and Mask for Lookup Table should also be tried.

10.11.3   Manipulating BLAST to Search Your Query Sequence Where You Want

10.11.3.1 Instructing BLAST to Ignore Specific Sequences within a Query (Masking). Both protein and nucleotide BLAST contain a "Filters and Masking" set of options in the Algorithm Parameters section of their respective BLAST pages (Figure 10.7.D). Two of these, "masking low complexity regions" and the "mask for lookup table only", are pre-configured for use by NCBI and can only be turned on or off by users. By default, they are turned off. Clicking on the blue question mark button will open up a detailed explanation of these options. If unsure of whether or not one or both these areas should be masked for any given Query sequence, treat BLAST as an experiment. Run BLAST testing and analyzing results for the effects of masking these regions. Use the Edit and Resubmit function (Section 10.10) of BLAST to perform these "experiments".

The third option, "Mask lower case letters", is customizable by users. Through this option, users can direct BLAST to ignore any stretch, or combination of stretches, of sequences within a query including deliberately masking any region of the query that seems to predominate in

BLAST hits. By masking sequences, the user is instructing BLAST not to return hits in masked regions, effectively weeding them from results. The area to be masked must first be determined. In this exercise that will be performed by running a preliminary BLAST search.

1. Go to the Protein BLAST home page (http://blast.ncbi.nlm.nih.gov/Blast.cgi) to begin a new protein BLAST search.
2. Reset the page and clear the query box as before.
3. In the query search box, type the following accession number: NP_571611.1.
4. Give a Job Title of: UnmaskedSequence.
5. Select 500 Maximum Target Sequences.
6. BLAST.
7. When the results draw, scroll down through the Descriptions section. You will note that most hits are in sequences containing a homeobox domain HOX, yet that domain is a small component of the query as seen in the Graphic Summary section. (The Query is 329 amino acids long, with the Homeobox domain restricted to only 50 amino acids between approximate amino acids position 230 to 280). To attempt to recover hits from the large majority of the query instead of only from the HOX domain, the sequence surrounding and in the area of the Homeobox can be "masked" from BLAST as shown in Figure 10.17.

   To mask the HOX domain from BLAST:
8. Determine where in the sequence the mask should occur from the results BLAST page. For this exercise a generous masking around the Homeobox domain was chosen, from residues 205 to 290.
9. Retrieve the sequence from NCBI for this exercise:
   a. Go to the NCBI home page (http://www.ncbi.nlm.nih.gov/), type "NP_57611.1" into the search box "All Databases" at the top of the page.
   b. Click the Search button.
   c. Select the Protein database which has a single hit, which is to the NP_57611.1 record.
   d. Once at the NP-57611.1 record, scroll all the way down to the bottom of the record and copy the numbered protein sequence.
10. Paste the numbered protein sequence into a Word processing program and convert to uppercase letters.
11. Locate sequence residues 205–290 and convert them to lower case letters as shown in Figure 10.16.

Unmasked Sequence 1:

1 MSTFLDFSSI SGGGDGGSGG SCSVRAFHGD HGLSTFQSSC AVRLNSCSGD ERFMSNISSQ

61 DVINSQPQQA GSYQSPGTLS ITYSAHPSYG TQSFCTGYNH YALNQDVESS VSFPQCGPLV

121 YSGNISSTVV QHRHHRHGYS SGNVHLHGQF QYGSATYGNS SDQANLTFVA GCSNPLSPLH

181 VPHHDACCSP LSDGVPTGQT FDWMKVKRNP PKTGKAGEYG FGGQPNTVRT NFSTKQLTEL

241 EKEFHFNKYL TRARRVEIAA SLQLNETQVK IWFQNRRMKQ KKREKEGLLP KSLSEQKDGL

301 EKTEDASEKS PSAPSTPSPS PTVEAYSSN

Masked Sequence 2:

1 MSTFLDFSSI SGGGDGGSGG SCSVRAFHGD HGLSTFQSSC AVRLNSCSGD ERFMSNISSQ

61 DVINSQPQQA GSYQSPGTLS ITYSAHPSYG TQSFCTGYNH YALNQDVESS VSFPQCGPLV

121 YSGNISSTVV QHRHHRHGYS SGNVHLHGQF QYGSATYGNS SDQANLTFVA GCSNPLSPLH

181 VPHHDACCSP LSDGVPTGQT FDWmkvkrnp pktgkageyg fggqpntvrt nfstkqltel

241 ekefhfnkyl trarrveiaa slqlnetqvk iwfqnrrmkq kkrekegllp KSLSEQKDGL

301 EKTEDASEKS PSAPSTPSPS PTVEAYSSN

Figure 10.17   Instruct BLAST to ignore portions of a query sequence through the use of the "mask lower case" option by converting that portion of the query to be masked to lower case letters and clicking on the "Mask Lower Case" box in the Algorithm Parameters section of BLAST.

12. Return to the unmasked search performed in step 7 above and click on Edit and Resubmit.

13. Reset the page, clear the search box and paste in the numbered "masked sequence".

14. Give it the Job Title: MaskedHOXSequence.

15. Select 500 Maximum Target Sequences.

16. *Finally check the "Mask lower Case" box at the very bottom of the Algorithm Parameters section (*Figure 10.7D*).*

17. BLAST.

EXPLANATION: The number of hits retrieved drop from 500+ in the unmasked search to 86 in the masked search. *BLAST will still recognize the presence of the HOX domain, but it will not return similarity hits originating in that area. Homeodomain hits may still be retrieved by BLAST if the subject sequence simultaneously has hits to the unmasked query portions AND contains a homeodomain.*

10.11.3.2   Instructing BLAST to Focus to Particular Sequences within a Query (Subsequence Searching).   Use the Query subrange function (Figure 10.6F) to focus BLAST similarity searches to specified subset of query residues. This results in a BLAST search opposite in concept to that of masking sequences. When a decision is made to mask sequences opting to also run a subrange search should be considered. Subrange searches are available at both protein and nucleotide BLAST.

1. From within protein BLAST from the above exercise, Click the Edit and Resubmit link to rerun a protein BLAST search.
2. Reset the page and clear the query box as before.
3. In the query search box, type the following accession record number: BAC42010.1.
4. Give the Job Title: SubRangeSearch.
5. Select 500 Maximum Target Sequences.
6. BLAST.
7. Reformat to show 500 records in the Graphical Summary.
8. When the results draw, a decision is made to focus a similarity search to the right side of the query, from residues 575 to 883, and away from the major TPR domain.
9. Click on the Edit and Resubmit link.
10. In the Query subrange box (Figure 10.6F) type in 575 in the "From" box and 883 in the "To" box,
11. BLAST.
12. The number of records retrieved drops to 107 with the results focused to finding similarity on the carboxyl terminal (query right side) end of the protein sequence.

REFERENCES

1. S. F. Altschul, W. Gish, W. Miller, E. W. Myers and D. J. Lipman, *J. Mol. Biol.*, 1990, **215**(2), 403.
2. S. F. Altschul, T. L. Madden, A. A. Schäffer, J. Zhang, Z. Zhang, W. Miller and D. J. Lipman, *Nucleic Acids Res.*, 1997, **25**(17), 3389.
3. S. B. Needleman and C. D. Wunsch, *J. Mol. Biol.*, 1970, **48**(3), 443.
4. T. F. Smith and M. S. Waterman, *J. Mol. Biol.*, 1981, **147**(1), 195.
5. D. J. Lipman and W. R. Pearson, *Science*, 1985, **227**(4693), 1435.
6. W. R. Pearson and D. J. Lipman, *Proc. Natl. Acad. Sci.*, 1988, **85**(8), 2444.
7. S. Pietrokovski, J. G. Henikoff and S. Henikoff, *Nucleic Acids. Res.*, 1966, **24**(1), 197.
8. D. Wheeler, *Curr. Protoc. Bioinformatics*, 2003, Chapter 3, Unit 3.5.1.

Subject Index